FIGHTING

FOR LOVE

IN THE

CENTURY OF

EXTINCTION

FIGHTING FOR LOVE IN THE CENTURY OF EXTINCTION

HOW PASSION AND POLITICS CAN STOP GLOBAL WARMING

EBAN GOODSTEIN

UNIVERSITY OF VERMONT PRESS
BURLINGTON, VERMONT

PUBLISHED BY UNIVERSITY PRESS OF NEW ENGLAND
HANOVER AND LONDON

UNIVERSITY OF VERMONT PRESS
Published by University Press of New England,
One Court Street, Lebanon, NH 03766
www.upne.com

Library of Congress Cataloging-in-Publication Data

Goodstein, Eban S., 1960–
Fighting for love in the century of extinction : how passion and politics can stop global warming / Eban Goodstein.
    p. cm.
Includes bibliographical references.
ISBN-13: 978-1-58465-657-9 (cloth : alk. paper)
ISBN-10: 1-58465-657-3 (cloth : alk. paper)
    1. Global warming—Prevention.   2. Greenhouse effect, Atmospheric—Prevention.   3. Environmental ethics.   4. Environmental policy.   I. Title.
QC981.8.G56G665 2007
363.738'74—dc22   2007016404

Selections from Chapter 5, "Politics," first appeared in "Climate Change: What the World Needs Now Is . . . Politics," WORLDwatch, 19-1, January/February 2006. Used with permission.

University Press of New England is a member of the Green Press Initiative. The paper used in this book meets their minimum requirement for recycled paper.

TO CHUNGIN CHUNG

CONTENTS

ACKNOWLEDGEMENTS

This book began in some quiet hours in the hills above Point Reyes: thanks to Peter Barnes and the Mesa Refuge for the time I spent there. My father Marvin Goodstein; always my best man, Scott Highleyman; Jon Isham and Ross Gelbspan, and all read and helped shape the book as it moved along. Thanks as well to Dallas Burtraw, Julian Dautremont-Smith, Matthew Follett, Judith Helfand, and Dylan Smith for helpful conversations. Phyllis Deutsch has been a wonderful editor, nudging me towards clarity. I am also grateful to the production team at UPNE.

This book was rescued by the vision of my partner, Chungin Chung. She read a languishing five chapters, and immediately saw the heart that was escaping me. The title is hers. That title, and her fierce insistence on an honest voice, drove me to revision the entire book, and bring it to completion.

FIGHTING

FOR LOVE

IN THE

CENTURY OF

EXTINCTION

CHAPTER 1

THE CENTURY
OF EXTINCTION

We stand at a unique moment in human history. Decisions that are ours to make over the next decade—to stabilize global warming pollution and invest in clean energy solutions—will have a profound impact not only on our lives and the lives of our children but indeed for every human being who will ever walk the face of the planet from now until the end of time.

Unchecked, global warming will kill more people than has pollution from any other industrial process. But global warming will also soon take its place as a primary driver of a wave of mass extinction that will sweep the planet this century. This is a book about the fate of many beautiful things of the Earth, the meaning of extinction, and also about a way through, toward a sustainable and prosperous future.

Realizations

In my early twenties, I was living and working in Anchorage, Alaska. In June 1983, friends and I headed out on a sea kayaking trip in Prince William Sound. We paddled west from the oil port in Valdez to the old military base at Whittier; with detours, a journey of about 120 miles. About ten days out, on a rare sunny day,

1

we found ourselves wandering up a deep, narrow fjord, with mountains rising steeply on either side. As the day progressed, we began to see more and more small icebergs. And then, to our delight, we saw that many of the icebergs were populated—it was pupping time for both the seals and the sea otters. On dozens of larger bergs, mother seals and their one or two babies were sound asleep in the hot sun. On others, more wary sea otter moms were hanging out with their pups, keeping a close eye on us. Soon we were gliding silently in the middle of this ice-choked nursery.

Our boat was so quiet that three times we were able to float right next to an iceberg cradling a family of sleeping seals, to the point that we literally could have reached out and touched them. As an illustration of how close we were, we took this picture, without the aid of a telephoto lens (fig. 1). This was an incredible day in my life. And I remember that night, by a fire on the beach, we told each other that this was a place we would someday bring our children.

I have two daughters now, and do fully intend to take them there. Perhaps they will be fortunate enough to experience our sense of wonder. There is some hurry. Those icebergs that provided a safe haven from bears and other predators for the seal and otter pups two and a half decades ago had calved off of tidewater glaciers at the mouth of the fjord. But today, as a consequence of human-induced global warming, glaciers are melting in Alaska, and where I live in Oregon, and indeed throughout the temperate world. By the year 2030, if present warming trends continue, there will be no glaciers at all in Glacier National Park in Montana. And some-time equally soon, there will be no ice for the seals and otters in Prince William Sound. This means—with a very high probability—that none of this life will be there for my daughters to show their children.

I first realized this connection between climate change and the looming extinction of marine mammals from parts of Prince William Sound in 1999, when I began speaking on college cam-puses about global warming. But, as I told and retold the story over the next three years, it was always as a hypothetical, a some-what distant, future event. So, one can imagine my profound sad-ness when in April of 2001, I read these words, tucked away in the back of the *Boston Globe:*

> MONTREAL—The early disappearance of ice in Canada's Gulf of St. Lawrence, which some scientists believe is linked to global warming, is wreaking havoc on the harp seals—which give birth on the floes—and causing economic hard-ship for hard-pressed fishermen who depend on the contro-versial spring hunt.
>
> Hundreds of drowned seal pups have already washed up on the shores of Newfoundland after their mothers gave birth in open water, apparently unable to find ice. The final death toll of the pups may be in the hundreds of thousands.

3

"The hundreds of thousands."

This is one small episode in a little-noticed, planetary holocaust: The twenty-first century is witnessing the unfolding of the sixth mass extinction since complex life emerged on Earth. The last of these mass extinctions occurred 65 million years ago; scientists believe that an asteroid approximately 10 kilometers across slammed into what is now the Yucatan Peninsula of Mexico. Dust from the cataclysmic impact, and soot from massive fires, led to rapid global cooling, which wiped out some 70 percent of life on Earth—including the dinosaurs—in the space of a few decades. Since then, life has recovered, thrived, and diversified into glorious complexity. Humans have inherited this bounty of 65 million years of largely uninterrupted evolution.

Today, life faces a threat of similar magnitude, resulting not from natural catastrophe, but from human-induced changes to the biosphere. Overall, biologists estimate that species extinction is proceeding now at a rate one to ten thousand times higher than the natural rate. For most of human history, overhunting was the primary cause of extinction, and for some creatures—commercial fish, primates—this remains true. Over the last few centuries, habitat destruction and the spread of invasive species have become the main extinction sources. With the passing of a few more decades, however, global warming will join these factors as a prime driver behind the worldwide dieback of diversity.

In 2004, scientists reported in the journal *Nature* that by 2050 habitat destruction from global warming alone could lock onto an extinction path up to a million terrestrial species. This is *35 percent of the estimated creatures and plants inhabiting the planet,* including, for example, more than one in five species of wildflowers. The authors conclude that anthropogenic warming is "likely to be the greatest threat [to biodiversity] in many, if not most regions." It is not hard to see why. According to the UN's Intergovernmental Panel on Climate Change, the Earth's average temperature is

4

likely to rise over the next century by somewhere between 2.5 and 10.5 degrees Fahrenheit. To put this range in perspective, during the last ice age—a period of time during which much of North America was covered with hundreds of meters of ice—the average temperature was only 9 degrees colder than it is today.

Within our children's lifetimes, we are thus looking at the very real possibility of a swing in global temperatures of ice-age magnitude, only in the opposite direction. Even the less extreme warming that is already locked in—3 to 4 degrees—will cause suffering for hundreds of millions of people across the planet, and devastation for many of our ecological systems. Because of the rapidity of the warming that we expect, throughout the rest of this book I will follow the scientist James Lovelock's recommendation and refer not to "global warming" but instead to "global heating." This dramatic rise in planetary temperatures demands that we think of the beauty at our feet as we last walked through a flower-graced meadow, take away 20 percent, then multiply the difference in that meadow by an entire planet. And of course, global heating will not come miraculously to a halt in 2050.

More stories.

Worldwide, there has been a recent rapid decline of frogs and other amphibians, punctuated by reports of widespread physical deformities, with dozens of species suddenly disappearing in western North America, Central America, northeast Australia, and Puerto Rico. Researchers believe that most of these declines, including many in undeveloped areas, have occurred in the last twenty-five years.

As the ice pack disappears in Canada's Hudson Bay, the polar bear population there is starving its way toward extinction. For the past twenty years, researchers have been monitoring the bears closely. The ice is breaking up two weeks earlier than it did at the beginning of the study period, making seal hunting increasingly difficult. The bears are coming back from the spring hunt an average of 22 pounds lighter. Weight loss is leading females to have fewer cubs and increasing mortality among the teenagers who are in line to replace breeding stocks. And every summer, it gets warmer.

In 1982, I worked for a mineral exploration crew in the Kalahari Desert in southern Africa. Another of life's vivid memories: blazing orange horizon, fading into an inky dark-blue night, and then, for hours, the deep, rumbling roars from a lion pride somewhere over the hill. But the lions too are almost gone. Since I stood in the desert, just twenty-five years ago, the number of wild lions has dropped 90 percent, from 200,000 to fewer than 23,000.

Two-thirds of the world's turtles and tortoises are threatened with extinction.

Chimpanzee populations have fallen from 2 million a century ago to 200,000 a decade ago, to fewer than 110,000 now. Along with gorillas and our charismatic near-relative the bonobo, failure to protect habitat is creating an increasing likelihood of near-term extinction for wild chimpanzees. According to Peter Walsh of Princeton University, "the populations are crashing really quickly. In the course of about ten years we're going to be in the situation where gorillas and chimpanzees are going to go from being widely distributed and abundant to being just in a few small pockets. That's bad because in the long term those small pockets just won't be sustainable, we won't be able to protect ten here and fifty there—it will be impossible. The idea that our closest relatives would go extinct is shattering to us. We don't want to have people looking back in twenty years and say they didn't do enough—they knew what to do but they didn't do it."

In 2004, my daughters and I took a trip to Hawaii. We were privileged to swim through the Kealakekua marine preserve, an amazing, pristine, coral forest rich with colorful inhabitants: lizard-fishes, snappers, eels, goatfishes, parrotfishes, angel, damsel, and surgeonfishes, emperors, and dozens and dozens more. Underwater, Kealakekua probably looks a lot like it did two hundred years ago, when the English explorer Captain Cook was killed on the neighboring beach. But across the planet, global heating has already damaged 15 percent of reefs beyond repair, and another 30 percent are at risk over the next thirty years. Studies predict that Australia's Great Barrier Reef, now one of the healthiest in the world, likely will be largely dead by 2050. A recent report on corals and climate change offered a feeble attempt at optimism: "While the net effects of climate change on coral reefs will be negative, coral reef organisms and communities are not necessarily doomed to total extinction."

It is stories like these that have brought us—reader and writer of this book—together. As the writer, I am, among other things, a father to my daughters, a son to my father and to the memory of my mother. I am a lover of nature, of creation. I am a one-time geologist, and now a teacher of environmental and natural resource economics at a college in Portland Oregon. I am a citizen of the United States of America. I am an active citizen. Over the last seven years, I have talked with thousands of people across the country about global heating, in schools, business clubs, churches and synagogues. I have organized over a dozen weekend workshops training several hundred volunteers to do the same. And now I am helping lead a national educational initiative on global heating called Focus the Nation. I have thought and listened and spoken about extinction a lot.

These extinction stories are not all about global heating, but collectively, they powerfully foreshadow the future that global heating could bring. In the year 2007, seals, otters, lions, turtles, frogs, apes, snakes, butterflies, polar bears, cheetahs, whales are disappearing along with their variously furnished homes: cloud forests, rain forests, ice pack, boreal forests, coral reefs, forests of deciduous trees, conifer, and palm. This is a very, very hard reality. What are we to make of this massive planetary dieback? What does it mean to us, Earth's sentient species, to see so many of our co-inhabitants, creatures with whom we have co-evolved over eons, disappear from the face of the world forever?

Reporting about endangered species and extinction typically follows a format to which we have all become accustomed: stories of dire ecological threat, followed by a good news preservation story. There is hope! the reporters feel obliged to add. But in the current world political context, no one believes it. When, rarely, the subject arises, I feel the same hopelessness that I see in the eyes of my friends. Why fight a war that already has been lost? To understand that this is the wrong question, we need to stare a difficult reality fully in the face. Regardless of our actions, ours will indeed be the century of extinction. And yet, even recognizing this, so much remains of the wealth and power of the natural world that is ours to pass on, so many beautiful gifts for the future.

People sometimes ask me: You talk about global heating and mass extinction all the time. Isn't that depressing? Doesn't that make you hopeless? And I do think there are two futures. The first is dystopic. In 2100, the Earth could be very hot, with temperatures up 10 degrees Fahrenheit on average. Combined with mass extinctions, post-peak-oil poverty, water shortages, and a reversion to tribalistic politics, all this might leave our descendants suffering. Or perhaps, a world rewired with clean-energy technologies, a globalization that raises the mass of humanity toward a decent standard of living, and a newfound and profound respect for the remains of creation will carry humanity forward into a new era of progress. I am not pessimistic about the future. I am not optimistic either. I am agnostic about what the world will really be like in fifty or one hundred years.

What I do know today is that I am very alive to this second possibility, the vision of a rich future for our children. When I walk in the mountains close to my home, I understand in ways that I did not used to, that much of the creation around me is disappearing, that one hundred years of steadily rising temperatures will elimi-nate the glaciers and the fish and insects and animals that their flowing streams and rivers support, and burn down many of the forests that blanket the slopes. In spite of this, each day I find that the life that surrounds me becomes more valuable, richer in the texture that it reveals about the universe, more precious in hidden promises of things yet unrevealed. And because of how I feel, I am grateful for the opportunity to fight for the things I love, to be part of this movement that can stop the heating of our world.

The next decade will be the most exciting, most decisive, most *human* time ever to be alive. We stand at a moment in history without precedent. Now, we hold in our hands truly incredible gifts for our children and our children's children: gifts of bright fish among sand and coral; gifts of intact glaciers and ice sheets the size of continents; gifts of polar bear and seals and salmon; gifts of frogs and ancient forests.

What amazing gifts. What a time to be alive.

Direction

To help ensure the survival of much of the beauty and bounty of creation, the critical thing we must do is to stabilize the climate and keep global heating as low as possible. There are two reasons for this. The first is that habitat destruction from global heating will soon join land conversion as a primary driver of extinction. Human-induced extinction has been with us for tens of thousands of years. Throughout prehistory, as human colonizers opened up new frontiers—in Australia, the Polynesian islands, and North America—many game species quickly disappeared. Over the past century, the human footprint has doubled, and redoubled, and re-doubled again. The number of people has grown from 1 to 6.3 bil-lion, and the average consumption level has quadrupled. In the face of this exponential growth in human impact, the pace of ex-tinction has accelerated almost out of control, as virtually no place on the planet is now immune from the age-old pressures of habitat conversion and hunting, or the newer force of invasive species.

Today, however, the overarching threat to species diversity is becoming global heating. Humans are now engaged in an unprecedented natural experiment, in which we are altering the fundamental nature of Earth's climate control system. What difference will several degrees of heating make? Consider how a human body responds if it heats up one or two degrees: It sickens. Three or four degrees warming, sustained, and the human dies. Ecosystems have evolved with a similar sensitivity to temperature. Increased heat, altered rainfall patterns, the spread of pests and diseases—all these factors threaten to vastly simplify natural ecosystems, eliminating delicate creatures and plants that have taken advantage of relative climate stability, and leaving behind more primitive and hardy colonizers. The official forecast for human-induced global heating within our grandchildren's lifetime is up to *10 degrees Fahrenheit.*

Over hundred-thousand-year ice-age cycles, in the past the planet has experienced swings in temperature greater than 10 degrees. Most species accommodated these gradual changes, migrating in the face of slowly advancing and retreating ice sheets. But today, the speed of global heating is dramatic. Even when northward migration of the ecosystems in which species function is possible, many systems will not be able to move that fast. And of course today, with many species only surviving in isolated pockets of habitat, fragmented by roads, fences, and human development, migration routes often do not exist.

If the climate must be stabilized to halt the direct destruction of species and ecosystems, stopping global heating is also our only real hope for a comprehensive solution to the other major global extinction driver: habitat destruction by humans, especially of ancient forests. Forests are home to the vast majority of the world's land-dwelling species; tropical forests alone account for more than half of these animals, plants, and insects. Rainforests hold special value: Although they cover only 6 percent of the world's landmass, they host more than half of the planet's known organisms. Forests also store massive amounts of carbon in their trees and soils. Preserving those forests will be key to preventing a devastating pulse of global-heating pollution from entering the atmosphere. Stopping global heating will thus require a global effort to prevent deforestation.

And we have a decade to take the first big steps. Writing in the journal *Science,* climatologists Brian O'Neil and Michael Oppenheimer have shown that without a serious pollution-reduction program in place soon, it will be exceedingly difficult to hold the planetary warming to the low, manageable end of the forecast—3 to 4 degrees Fahrenheit or so. As we push the Earth beyond this limit, catastrophic outcomes, such as the collapse of the West Antarctic or Greenland ice sheets, a Gulf Stream disruption, or the massive release of methane trapped in permafrost and on the seabed floor, become more and more likely. The top U.S. government climate scientist, Dr. James Hansen of NASA, speaks in very plain language:

> How far can it go? The last time the world was three degrees [C, or 6 degrees F] warmer than today—which is what we expect later this century—sea levels were 25 meters higher. So that is what we can look forward to if we don't act soon . . . How long have we got? We have to stabilize emissions of carbon dioxide *within a decade,* or temperatures will warm by more than one degree. That will be warmer than it has been for half a million years, and many things could become unstoppable. If we are to stop that, we cannot wait for new technologies like capturing emissions from burning coal. We have to act with what we have. This decade, that means focusing on energy efficiency and re-

newable sources of energy that do not burn carbon. *We don't have much time left.* [Emphasis added]

This is a paragraph that every American should read, and read again. The top U.S. government climate scientist has told us: "We have to stabilize emissions of carbon dioxide *within a decade. . . We don't have much time left.*"

Global heating can be stopped, but only through a massive effort to rewire the planet with safe, affordable, clean energy: electricity produced from renewable sources like wind and solar. At the same time, we can also engineer vehicles that run on biofuels, or hydrogen, or batteries, using power from these clean electricity sources. Chapter 6 of this book focuses on solutions. The obstacle to building a prosperous future through a clean energy revolution—and the millions of new jobs that will come with it—is neither economic nor technical. Instead, the obstacle is a lack of political will. When we find that will to jump-start the new economy, then, to finally stabilize the climate, this wave of technological innovation can and must be coupled with a successful program to protect the vast sink of carbon tied up in the world's remaining ancient forests.

A clean-energy revolution. A massive effort to protect forests. Stabilizing the climate in this way is, without question, a political project of unprecedented scale. Yet the bottom line for life on the planet is also clear: To have any serious impact on the extinction wave that is sweeping the planet, the government of the United States must, soon, play a very, very active role in shaping the trajectory that global markets—especially in energy—will follow. Leadership must come from America, from our federal government, in part because we are by far the world's biggest global-heating polluter, but more fundamentally because we have the resources and the ingenuity to develop rapidly and market the clean energy technologies that can break the global dependence on fossil fuels.

Clearly, a successful effort to stop global heating will require bold political vision. In Europe, Japan, Canada, and in some of the states on the West Coast and in the Northeast political leadership—both Republican and Democratic—is beginning to move in that direction. But the progress is slow and halting. At the national level, there has been no action on global heating for the last seven years.

To hold global heating to the low end of a few degrees, change must happen, and very quickly. The consequences of continued inaction will mean tremendous human suffering and a disappearance of many, many of the creatures of the Earth. But will these concerns be sufficient to drive Americans to the kind of action nature is demanding? Many of us are beginning to understand the devastating impacts on the lives of people from unchecked global heating; a recent U.K. government report is predicting potential economic consequences on the scale of a great depression or world war. But what of the rest of the life on the planet? Over the past three decades, Americans have grappled with extinction on a case-by-case basis: a snail darter, a bald eagle, a spotted owl. What is the meaning of the coming wave of mass extinction, the permanent loss of hundreds of thousands of species?

Wealth, Knowledge and Spirit

Where I live in the Northwest, many people love wild salmon. These brilliant animals are cultural symbols and spiritual metaphors. We are fascinated by their life journey: as young smolts, to sea, and as mature adults, home again. They travel, unerringly, hundreds of miles against strong currents, over obstacles nature- and human-made, to breed, to die, to fertilize the forests that nourish life.

Wild salmon fisheries in the Pacific Northwest have collapsed and many are on the verge of extinction. The reasons for this are complex, but clearcut logging has played an important role. In an old-growth forest, fallen conifers create dams that in turn develop spawning pools; different types of hardwood trees—maple and later alder—provide food for insects upon which young salmon smolts feed. Large trees throughout the watershed protect streams from devastating debris flows during heavy rains. Slowly, these trees have been removed and replaced by a much less diverse second-growth forest, destroying the resilience of the ecosystem. Now, catastrophic events like mud flows, once a part of stream rejuvenation, scour and destroy the salmon habitat.

What are wild salmon worth?

The direct economic value of salmon, it turns out, is relatively small. On the commercial front, even as wild salmon populations have plummeted, global supplies of the fish have expanded exponentially. In the United States, salmon in grocery stores is cheaper and more widely available then ever. The source is salmon farms. While salmon aquaculture currently depends on unsustainable harvesting of ocean fish for feed, nevertheless, wild fish from the Northwest can no longer compete as a serious source of food. Recreational benefits are larger: Salmon generate income for businesses related to anglers and ecotourists. But one can overstate the importance of salmon as a driving force behind outdoor recreation. As these fish disappear, new species will populate the streams and rivers, and recreational industries will adapt. Wild salmon might possess potentially large economic value, hidden away on their genetic code. As global aquaculture continues to expand, the genetic diversity in populations scattered across the Northwest might be very important for breeding meatier, tastier, or more disease-resistant populations.

Moving from the direct economic value of the fish, salmon inhabit communities of interwoven species, communities that can yield "ecosystem services" such as water filtration and atmospheric regulation. I sometimes press my students on the ecosystem value of salmon; they typically argue that bears will be deprived of food (although we have very few left); aquatic habitat and nutrient cycles will be disrupted; and more generally, the "web of life" will be damaged. Young people have a healthy sense that everything is connected to everything in some strong way. They also have a strong belief—though one that is less empirically well grounded— that the connection runs primarily through human material welfare. One more rip in the ecological fabric, so the thinking goes, will lead to some very scary, though often unspecific, catastrophe for our economic and social system.

And yet, humans have been tearing at the web of life for a long, long time, and there is good reason to believe that, from a practical point of view, humans can easily survive dramatic alterations—and simplifications—of local ecological systems. Midwestern prairies have disappeared, New England hardwood forests have been decimated and have re-emerged in a less diverse form, Western rivers have been dammed and reduced to a trickle, Northwestern salmon populations have fallen dramatically, and human life goes on.

There is a third level of value. Salmon and the ecosystems in which they thrive are the products of millions of years of natural engineering. The most incredible feature of salmon is their homing instinct, apparently driven by the unique "smell" of the stream in which they were born. We do not understand the details of how this works. Spiders spin webs of organic fiber stronger than anything we can engineer. We do not understand the details of how this works. Geckos hang upside down on plateglass, held as if by magic. Three years ago, we began finally to understand how this works (more on geckos later). Every creature that we drive to extinction may hold the promise of a unique engineering solution to important human problems that, once unlocked, could vastly enrich our descendants.

Chapters 2 and 3 of this book ("Wealth" and "Knowledge") explore the question of whether the scale of mass extinction we are witnessing now, in the twenty-first century, has more ominous implications for human welfare, in strictly practical terms, than it did in the twentieth. By destroying so much life, are we destroying as well the foundation of our own material prosperity? By tearing out so many pages in the book of life, are we depriving future generations of a vast store of knowledge?

There is a final value to salmon.

My students are shy about raising moral issues related to salmon extinction, perhaps because they are in an economics class, and they expect me to expect them to stay in that box for ninety minutes. They often do point out that salmon are culturally significant to Northwest Indians, but they do not easily articulate a spiritual relationship to the creatures of their own. Last year, after having students list on the board the economic costs of salmon extinction, I pressed a student on the issue:

> ME: Is it wrong to drive salmon to extinction?
>
> BARBARA: It would be sad.
>
> ME: But is it wrong?
>
> BARBARA: It would be sad.
>
> ME: Would it be morally, ethically wrong?
>
> BARBARA: It would be sad.
>
> ME: Okay, but would it be wrong?
>
> BARBARA: It would be sad.

It was, in fact, this exchange that motivated me to write this book. There is no hope in sadness. I knew this young woman to have a deep connection to the natural world, but she could not claim a public, moral language to express feelings of either awe or anger.

This in itself is sad, and is reflective of the direction that environmental discourse has headed over the last half-century. People who love nature have a very hard time simply saying that. Instead, justifications for preserving species must be articulated in "greater good" language. And, of course, those economic and knowledge arguments are important. But the inner fire driving people to preserve and protect creatures and plants across the globe is seldom explicitly acknowledged or nurtured.

This has not always been so. John Muir—whose father was an evangelical preacher—spoke eloquently and easily of the transcendental power of life and of the holy face of nature. His recent biographer, Dennis Williams, argues that Muir was driven by a vision of the wilderness as a pure place, a place so illustrative of God's grace that it had redemptive power for sinful man. Here is Muir:

> We seem to imagine that since Herod beheaded John the
> Baptist there is no longer any voice crying in the wilderness.
> But no one in the wilderness can possibly make such a mis-
> take. No wilderness in the world is so desolate as to be with-
> out God's ministers. The love of God covers all the earth as
> the sky covers it and fills in every pore. And this love has
> voices heard by all who have ears to hear. Everything breaks
> into songs of Divine Love just as banks of snow, cold and
> silent, burst forth into songful, cascading water. Yosemite
> Creek is at once one of the most sublime and sweetest
> voiced evangels of the Wilderness of the Sierra.

It was this religious worldview that was the ultimate passion behind his tireless writing and advocacy for wilderness preservation.

There are religious people fighting global heating who still come at their work from this framework. But where does this leave those, who, like the majority of my students are "spiritual but not religious" or agnostic, and thus have no natural moral language? Even worse, what is an atheist to do? A deeply felt love of nature and its diversity motivates many people's concern about climate destablization, and yet too often, we can only speak out loud of the abundance of life using the deadening jargon of public policy.

Ultimately, there is mystery at the heart of a worldview that embraces a love of creation. The most rational defender of nature is driven by a passion for wildness that cannot be explained by an appeal to logic. That passion can begin to be understood, however, at least partially, by recognizing that people are sentient creatures who evolved as hunter-gatherers, immersed in the natural world, for several million years. Our psychology—to the extent that it is a product of the hard-wiring of evolution—has for some reason equipped us with the awe-inspired capacity to love creation in all of its diversity. Whether springing from Nature or nature's God, this unique human capability is the only thing that can drive us to slow the ongoing rush to mass extinction.

The human community needs, above all, a respiritualized language that acknowledges this love, to carry it forward in this century of extinction. Stabilizing the climate demands political will. And without effective moral language, real politics is impossible. Chapter 4 ("Spirit") explores the most important and least acknowledged value of the salmon, and of creation as we have inherited it: Our love for them.

Politics

It is a premise of this book that if we are to hold global heating to the manageable, low end, then very soon, within the next ten years, America must do two things: Freeze emissions of global-heating pollutants, and begin to invest tens of billions of dollars every year in the clean-energy technology solutions that our kids will need in 2030, when the planet's temperature is rising fast. In the late 1990s, America seemed to be on the road to action. We had signed the Kyoto global-warming accord, and candidate Al Gore was committed to the treaty. Candidate George W. Bush, largely silent on Kyoto, nevertheless pledged to regulate carbon dioxide—the main global-heating gas—as a pollutant. And yet, America's apparent commitment to action on global heating was in fact quite hollow. Once in office, President Bush promptly reneged on his promise to regulate CO_2, and then defiantly pulled the U.S. out of the Kyoto process, even as Europe, Canada, Japan, and Russia moved forward to ratify the treaty. In doing so, Bush reflected a new constellation of power in American politics.

Over the last twenty-five years, a new and rigid political ideology came to power in Washington; an ideology that believes, above all, that "government is the problem." Outside of the Northeast, an older generation of moderate, pragmatic Republicans—with a long tradition of strong support for environmental protection—largely disappeared. In their place came a new breed of politician who pledges strict allegiance to a very old way of thinking, a laissez-faire economics reminiscent of the 1890s. With an electoral base of social conservatives, this new political establishment is supported by a string of think tanks and a self-consciously right-wing media, including radio talk shows, FOX news, the *Washington Times,* and the *Wall Street Journal.*

From 2000 to 2007, "government is the problem" politicians dominated the leadership of the Republican Party, and controlled the federal government: the Presidency, and for much of the time, both houses of Congress. This group of politicians is deeply opposed to the very idea of ambitious government policy, and, sadly, has proven quite willing to distort inconvenient scientific facts that suggest any need for serious action. Since activist government is the beast they are sworn to slay, under their watch, we will simply never see more than token climate policy put in place.

The 2006 elections, a referendum on the war in Iraq, also had the effect of ending America's experiment with "government is the problem" leadership in the House, and—just barely—the Senate. The elections also exposed deep rifts in the Republican Party, creating space for a new leadership that can embrace the GOP's older tradition of support for protecting the natural world. And yet the "government is the problem" ideology remains deeply entrenched within the Beltway and its adherents will continue to fight ambitious global warming policy at every turn.

Stabilizing the climate, therefore, will require strong leadership backed by increasingly solid clean-energy majorities in both the executive and legislative branches of government. Stabilizing the climate requires, fundamentally, the creation of a new, bipartisan governing coalition fully supportive of a sweeping global transformation of markets in energy and forest products. This kind of coalition can only be built through electoral politics. Electing clean-energy leaders into the Senate, the House, and the Oval Office— and getting it done in the next few years—is the *only real solution* to climate stabilization at acceptable levels, and by extension, the looming extinction crisis.

Global heating is not a partisan issue, but it demands, at its heart, ambitious action by the federal government. We can achieve a new national consensus around support for a clean-energy future first, if the current debate in the Republican Party over global heating can return the party away from its recent, inflexible, "government is the problem" mindset, to its historical pragmatic and "pro-creation" roots. At the same time, the Democrats will need to summon unusual courage, providing truly visionary leadership. And there is an important lesson from history to be taken here.

One hundred years ago, riding the back of the Progressive Movement, Republican President Teddy Roosevelt pursued a pragmatic, bipartisan course of reform. Challenging the antigovernment ideology of the large "Trusts," Roosevelt's generation pushed for health and safety regulation, democratic reforms like women's suffrage, and the creation of a vast conservation network of government-managed national forests and parks that is the envy of the world.

For the last decade, Washington has been, and remains, in the grip of a paralyzing antigovernment ideology, serving the interests of a few large corporations and a few of America's most rich and powerful people. But this time, the threat to the well-being of the nation—and to many, many creatures of the Earth—comes from a seriously provoked Nature. Unchecked, the impact of global heating will be catastrophic for people and ecosystems across the globe, and yet the "government is the problem" political establishment that is now dominant in the Republican Party, the political establishment attracting the support of close to half of America's voters, remains in deep ideological denial.

No generation before us has faced a decade of choices that will so profoundly impact the course of life on this planet as those we now face. And no generation before us has had the opportunity to enrich the future so vastly. If we seek to find any real meaning in the life that is fast disappearing around us, then we face a political impasse that must be broken. Both the Republican and the Democratic parties have proud traditions to recover, traditions on which we can build a twenty-first-century movement that once again harnesses the power of government to do good things. A new politics—and only a new politics—can spark the clean-energy revolution that will serve as a foundation for a new era of human prosperity, protect the world's forests, stabilize the climate, and preserve the diversity of life on the planet.

Forests

I grew up in a little town in the Cumberland Mountains of south central Tennessee. When I was a boy, every spring in May, the mountain was covered in a glorious white carpet of dogwood blossoms. We had a sleeping porch looking out over our back yard, and I used to sit up there in the evening and just drink in the setting, the smell, the gentle presence of the trees. A few years ago, after a twenty-year absence from dogwood season, I went home looking forward to seeing them. But when I arrived, there were no blossoms—in fact, the trees in my yard had no leaves. I thought maybe I got the time of year wrong. I asked a friend, and he said, "Oh no, the dogwoods are all dying. Killed by a blight." And I said "They are all dead?" And he said "Yes, mostly dead."

Dogwood blight burned through the southeastern United States in the space of about ten years—from the early 1980s to the 1990s, killing most of the trees in my part of the country. The source of the virus is not well understood; it is believed to be an invasive. We do know, however, that in the coming decades, one of the certain outcomes of human-induced climate change will be the rapid spread of new diseases and pests. Indeed, on my most recent visit home, I saw pine forests being cleared following a deadly infestation of beetles. A recent string of warmer winters had allowed the insects to spread north into the mountains. A similar infestation, one signature of global heating, is killing millions of acres of Alaskan spruce forest and the ponderosa pines in the Southwest.

Major climate models are predicting that if the Southeast warms up and dries up too much from global-heating pollution, the hardwood forests will disappear under the combined impact of heat, fire, and disease. According to these models, over the next century, major diebacks of oak, hickory, sycamore, poplar, and maple, will convert the deep, shaggy coves of the Southern mountains, into something that looks more like a savannah. When I look at those maps and see the Southeast turning from green to brown, it makes my heart sick. As a boy, I would never have thought that those dogwoods, which had graced the southern Cumberland Mountains for thousands of years, could largely disappear in the span of a decade. Well, they did. I have seen it.

I loved the Southern forest as it was, and I love it now, as it remains. And in this decade, during which the fate of much life in this world will be decided, I am grateful to be part of a fight for a new kind of politics, a politics that can lay the foundation for a clean-energy revolution, create a just and prosperous future, and pass on to all human generations to follow the beauty and the wealth of creation.

CHAPTER 2

WEALTH

Choose one of your most cherished wild creatures.

Hold an image of it in your mind.

Now ask yourself—why do I care, really, if the species lives or dies?

In your answer, you may have placed spiritual or moral concerns at the top. However, in this chapter, I want to focus on biodiversity as a source of material wealth. It is very likely that your plant or animal has some element of use value to humans: as food or medicine, for wildlife viewing, as a repository of genetic code with potential agricultural, industrial, or pharmaceutical value. These are economic roles at the *species* level.

Your creature undoubtedly also serves as a link in a broader *ecosystem* that generates a flow of value for people. This latter category encompasses economic functions that most of us seldom think about, since we take them so much for granted. In a path-breaking book called *Nature's Services*, the editor Gretchen Daily provides a partial list of the useful systemic functions of nature:

1. Purification of air and water
2. Mitigation of floods and droughts
3. Detoxification and decomposition of wastes
4. Generation and renewal of soil and soil fertility

5. Control of the vast majority of agricultural pests
6. Dispersal of seeds and translocation of nutrients
7. Maintenance of biodiversity, from which humanity has derived key elements of its agricultural and industrial enterprise
8. Protection from the sun's harmful ultraviolet rays
9. Partial stabilization of climate
10. Moderation of temperature extremes and the force of winds and waves
11. Support of diverse human cultures
12. Providing aesthetic beauty and intellectual stimulation that lift the human spirit

This chapter will focus on the meaning of life, both in its species-level function as a direct resource, and as a member of an ecosystem generating the services numbered 1 through 10 above: nature acting as a store of wealth generating a flow of directly practical values. We will defer a discussion of items 11 and 12, the support of cultural diversity, and aesthetic beauty and intellectual stimulation, to the next two chapters.

In her review of the economic benefits of natural systems, Daily argues "Failure to foster the continued delivery of ecosystem services undermines economic prosperity, forecloses options, and diminishes other aspects of human well-being; *it also threatens the very persistence of civilization*" [emphasis added]. Loss of biodiversity generally weakens the ability of nature to deliver ecosystem services: loss of biodiversity means loss of wealth. But is it a fatal loss? Is the mass extinction underway today seriously compromising the material welfare of future generations? Or will it merely rule out some options for a global population that, on average and in general, will continue to experience higher material standards of living: longer lives, improved medical care, more secure access to food, reduced infant and childhood mortality, better education, superior housing?

24

Looking Backward, Looking Forward

In 1930, a year into the Great Depression, the famous economist John Maynard Keynes wrote a surprisingly sunny little essay called "Economic Possibilities for Our Grandchildren." Surveying a century and a half of explosive economic growth in Europe and America, Keynes saw no reason to suppose that the underlying capitalist dynamic was about to grind to a halt. He concluded, with emphasis, "in the long run *that mankind is solving its economic problem,*" predicting "that the standard of living in progressive countries one hundred years hence will be between four and eight times as high as it is today." Seventy-five years on, Keynes prediction is on track to being realized; average incomes in Britain and the United States are about three and a half times higher than at the peak of the 1929 business cycle. In recent decades, rising inequality, especially in the United States, means that a typical family's material welfare is improving less rapidly than these per capita figures suggest. Still, periods of sustained economic growth such as the 1960s and 1990s saw broad material gains across all income groups.

The developing world too saw large per capita income increases from 1930 through the late 1970s. This progress has continued throughout much of East and South Asia, but largely stalled in Latin America, and actually reversed in parts of Africa. Today, the average standard of living in the middle-income countries approximates that of the United States and Britain in 1929. At the same time, it is important to recognize that economic gains in poor countries have come against a backdrop of explosive population growth: World population grew from 2 billion in 1929 to more than 6.5 billion today. While the number of people on the globe continues to climb, perhaps the best ecological news of the past fifteen years has been a rapid and unexpected decline in population growth rates. The UN's median estimate of future global population has fallen from a 1992 projection of stabilization at 12 billion by the end of the twenty-first century, to today's estimate of stabilization at under 10 billion.

In 2005, surveying the economic possibilities for *our* grandchildren, it is hard to share Keynes's boundless optimism. Keynes's progressive worldview was a modern one, essentially shaped before the War to End All Wars, and not after the Holocaust, Soviet gulags, Hiroshima, and Rwanda. In our postmodern era, faith in universal progress must necessarily be tempered. The twenty-first century undoubtedly will see periods of regional warfare, deployment of horrible new weapons, episodes of genocide, new diseases and epidemics, local economic collapse and stagnation, the occasional global depression, and continued poverty, hunger, and despair for hundreds of millions and probably billions of people.

But barring unpleasant surprises, it also seems likely that the relentless underlying march of material progress will continue. Assuming as well that encouraging trends toward declining fertility continue, then our grandchildren may indeed be living in a world where median global living standards rise to levels found in the 1960s in the rich countries. If the past is a good guide to the future, than one hundred years hence, Keynes's dream of a post-scarcity world will have come significantly closer to reality.

It may be, however, that that history is not a good guide. The economist Herman Daly points out that twentieth-century growth took place in an "empty world"; he argues that the economics of a "full world" are different. Recent increases in Gross Domestic Product (GDP) per capita disguise a whole gamut of costs associated with growth, including more pollution and congestion, and an unsustainable depletion of natural capital. To illustrate, we now face two very serious and closely related resource shortages: world supplies of cheap petroleum and—most critically—the ability of the atmosphere to absorb carbon dioxide without triggering catastrophic global heating.

26

For the last hundred years, both oil production and oil consumption have grown geometrically, in the last thirty years alone almost doubling, from levels of 45 million barrels per day to over 80 million. And as living standards in China, India, and other developing countries improve, demand growth shows no sign of slacking off. But supplies of cheap oil are limited. Sometime over the next few decades—maybe in five years, maybe in twenty-five— production of cheap oil must slow and then level off. But demand growth will not, unless affordable substitute technologies can be brought on line rather quickly. As global demand for such a critical commodity surges over a suddenly limited supply, some analysts foresee widescale chaos and conflict as the inevitable result.

There is a way to sidestep this "peak oil" problem: large-scale government investment in renewable electricity sources such as wind and solar. These energy sources could power the production of hydrogen and biomass fuels for our vehicles without the dramatic increase in pollution that would occur if we were forced to turn to coal to manufacture vehicle fuels when petroleum becomes too expensive to burn. Government needs to promote renewable electricity because the market alone simply "can't see" the peak-oil problem coming. The underlying problem is that petroleum reserve levels are not public information; the big oil companies and the Middle Eastern governments may have some sense of when their own peak-oil crunch is likely to come, but no one has the full picture. Because the precise date of the peak-oil crunch is uncertain, the private sector simply will not make the large-scale investments needed to manage the transition. But without these investments in renewable power sources, rapid, largely unanticipated increases in oil prices could severely damage the global economic system, could lead to widespread political unrest, and undoubtedly would spark a rush to an environmentally disastrous power alternative: coal.

And then, of course, there is global heating. Figure 2 is a simple

GLOBAL HEATING FOR NOBEL LAUREATES

refresher course on the subject: I put this picture up when I am talking about global heating, regardless of audience—sixth graders to graduate students.

The left-hand side of the picture shows how light from the Sun strikes the Earth and is reflected back as heat. Fortunately, the Earth is wrapped by a thin but cozy blanket that we call our atmosphere; the blanket traps heat from the Sun and keeps us warm. Mars has no atmosphere; it is too cold. Venus has too thick an atmosphere; it is too hot. But just like in Goldilocks and the Three Bears, until recently the thickness of the Earth's atmosphere has been *just right,* keeping temperatures and the climate stable.

One important component of the gaseous blanket of the atmosphere is carbon dioxide, or CO_2. Until 1870, for the last half a million years, the level of carbon dioxide had never risen above 290 parts per million. But as the right-hand side of figure 2 illustrates, due to both the burning of fossil fuels—oil, coal, and natural gas—and massive deforestation, the atmospheric concentration of CO_2 is increasing. It is now 382 parts per million, and rising at 1 to 2 parts per million per year.

Thicker blanket, warmer planet. This is the simple science of global heating, supporting a firm consensus in the scientific community that the planet is heating up. Over the last three decades, thousands of climate scientists from dozens of countries have turned the data inside out, trying to find some other possible explanation for the dramatic warming of the last century. The clear conclusion is that what we are living through is not a natural cycle. The only reasonable explanation for global heating is the simplest one: Human pollution is creating a carbon blanket that is beginning to smother the planet.

The consequences of a hotter planet range from the manageable to the catastrophic. Especially if the more extreme heating scenarios materialize, climate change has the potential to impoverish large portions of the planet, particularly the nations of the southern hemisphere. Poor people in developing countries will suffer most (and are already suffering most) from global heating, partly because they already tend to live in warmer and dryer parts of the world, but more fundamentally because they have far fewer resources with which to adapt to a rapidly changing climate.

Where I live, just a few degrees of warming is predicted to cut snow pack in the Oregon's Cascade Range by more than half by mid-century, and by the end of the century there could well be no snow at all in the mountains in early summer. This does not just mean bad skiing (although it does mean that). The snow pack is a gift from nature, storing hundreds of millions of acre-feet of water

29

to feed our summer streams and rivers. Snow pack loss will lead to dramatic reductions in summer water-supply, and we need that water, for irrigation, hydropower, to support salmon, and for recreation. Summer water prices will skyrocket, driving many farmers out of business. But the Northwest economy will not collapse—we can and must adapt to moderate climate change. Through conservation, construction of aquifer storage systems, and the utilization of new technology, we will figure out how to get by with vastly reduced summer water supplies. We will be poorer, our quality of life will fall, but we will adapt.

Further south in the western mountain chain of the Americas, however, the forecast is much grimmer. The inhabitants of Lima, Peru, depend on the snow-fed Rimac River for their water supply. Within a few decades, that river may well be largely dry half of the year, leaving 6.5 million people without water. It is very hard to imagine how these people will adapt. By 2050, a couple of billion people worldwide—mostly in India and China—will have their water supplies affected by global heating.

The scenario just described assumes a gradual warming. But increasingly, scientists have become worried about the possibility, perhaps occurring very soon, that our warming planet might cross some emission threshold, locking in some temperature increase which would lead to catastrophic outcomes. "Abrupt climate change," caused for example, by massive methane releases from thawing permafrost or a temperature-destabilized ocean floor, or by the disruption of the ocean's thermohaline circulation system (including the Gulf Stream), or by a fire-driven deforestation of the Amazon, could lead to sudden, large shifts in temperature, wreaking havoc in rich and poor countries alike. Or, we could set in motion the collapse of the West Antarctic or Greenland ice sheets, processes that ultimately would raise sea level by 40 feet, inundating all the world's coastal cities. These are all very scary—and very possible— scenarios with which John Maynard Keynes did not have to contend.

Running out of cheap oil and running out of atmosphere: The resource constraints of a "full world" may render economic progress in the twenty-first century much more difficult to achieve than it has been in the past. Still, over the last two hundred and fifty years, capitalism has proven to be an incredibly dynamic and adaptable economic system: Hope resides in the fact that, in our globalized economy, clean-energy technologies, once developed, could spread as quickly as computers and cell-phones. We may yet muddle through.

Here is how Keynes concluded his 1930 essay:

> The pace at which we can reach our destination of economic bliss will be governed by four things—our power to control our population, our determination to avoid wars and civil dissensions, our willingness to entrust to science the direction of those matters which are properly the concern of science, and the rate of accumulation as fixed by the margin between our production and our consumption; of which the last will easily look after itself, given the first three.

Keynes's point was that getting the politics right is what matters. If that can be done, then economic growth will largely take care of itself. This remains true, and in a twenty-first-century context, good governance is more critical than ever. Global, national, and regional political systems will have to continue to develop institutions to help stabilize population growth. Science, aggressively supported by government, must rapidly develop clean-energy alternatives that can sidestep a peak-oil crisis and enable the dramatic cuts in global-heating pollution needed to stabilize the climate. In spite of the magnitude of these challenges, they are not, currently, unmanageable. But the danger is clear. In the absence of visionary political leadership, these serious ecological constraints to reducing global poverty and promoting balanced economic progress will remain unacknowledged, and in that case, they may well become unmanageable.

This returns us to Gretchen Daily's point. Ecological limits bound our economic system. If left unaddressed, these could possibly "threaten the very persistence of civilization," and certainly lead to large-scale reductions in quality of life for a large percentage of the human population. Is biodiversity loss—like oil or climate stability—one of these pivotal resource constraints?

Economic Sustainability and Ecosystem Diversity

Keynes's progressive global vision for the grandchildren in recent decades has adopted a new name: "sustainable development." The term emerged on the world stage just twenty years ago, in the 1987 Bruntland Commission report for the United Nations Environment Programme (UNEP), entitled *Our Common Future.* The commission offered this definition: "The ability of humanity to ensure that it meets the needs of the present without compromising the ability of future generations to meet their own needs. Sustainable development is not a fixed state of harmony, but rather a process of change in which the exploitation of resources, the direction of investments, the orientation of technological development and institutional changes are made consistent with future as well as present needs."

Keying off this definition, "sustainability" as economists have come to understand it is a property of an economy by which the material welfare of the typical member of a society does not decline over time. Also in economic jargon, "biodiversity" is a resource to be exploited to promote material progress. Under the rubric of sustainability, reductions in biodiversity can only be justified if, *on net,* the process leads to improvement in the welfare of future generations. Over the last two centuries, Americans have indeed sacrificed genetic diversity for the sake of material progress— progress that does appear to have been largely "sustainable," based on this economic definition. New England's hardwood forests were largely clearcut to support the early American economy. The virtual destruction of the prairie ecosystems of the Midwestern United States have generated a vast, reliable, and valuable flow of crops, at least for the last 150 years. Northwestern hydroelectric dams, in spite of their devastating impact on salmon, electrified the region, almost certainly forestalling the construction of dozens of pollution-intensive, coal-fired power plants. While some of the dams may now have outlived their usefulness, they nevertheless laid the foundation for rising prosperity in the Pacific Northwest.

But these examples again beg the question of whether the past is a good guide to the future. It is obvious that humans cannot survive without the free services of nature. To underscore this point, in 1997, an interdisciplinary team of researchers published an article in *Science* that was a first, and extraordinarily ambitious, attempt to monetize the global value of ecosystem services. Using a variety of economic techniques for valuing nonmarket services, and extrapolating from local and regional studies, the study concluded that nature provides an income flow equivalent to $33 trillion annually. In the year of the study, global GDP was about half that size, at $17 trillion.

Ignoring the myriad technical issues associated with this kind of valuation exercise, the rhetorical point of the study is well taken: The total economic value of functioning ecosystems is *very big*. (One economist quipped that if the *Science* team was interested in really estimating the total value flow of natural services, then $33 trillion was "a serious underestimate of infinity.") Indeed, if the extinction crisis leads to a major compromise of the ability of ecosystems to deliver these services, then clearly the material welfare of humanity would be at risk. The difficult question to resolve is whether major ecosystem services—water purification, climate regulation, pest control, and so on—will be sustained adequately should 30 percent or more of Earth's species disappear. Before addressing this big issue, however, we can look at a more manageable piece. On the margin—at the level of individual resource development decisions—do the economic benefits of preserving ecosystems and their biological diversity exceed the costs?

Over the last decade, economists, sometimes working with ecologists, have begun to examine the benefits of preserving natural ecosystems, and weighing these benefits against the costs of failing to convert these lands to agricultural or other uses. Perhaps the most celebrated example lies in the Catskill Mountains, up the Hudson River from New York City. Much of New York's water comes from this region, and as the rural area has developed in recent years, increased run-off from roads, farms, and septic systems has lead to a decline in water quality. New York faced a choice: Build a $6- to $8-billion water treatment plant, or protect the watershed itself. Here the choice was an easy one. Conservation cost only $1 to $1.5 billion. As a side benefit of maintaining rural, forested land upstream of the city, New York also purchased cheap flood insurance.

I was witness to this particular benefit of controlling urban sprawl in my home town of Portland, Oregon, where the city has had in place an urban growth boundary since 1979. The boundary imposes very tight controls on development in the Willamette River Valley farmland upstream of the city, and has been quite effective in controlling suburban encroachment. Between its inception and 1995, Portland added only 5 square miles of development, while Denver, which had a similar population growth rate, expanded an astounding 180 square miles over a somewhat longer period.

In February 1996, the west side of Oregon, including the area around Portland, was ravaged by massive flooding, and downtown residents and business owners watched helplessly as the river rose higher and higher. The mayor ordered temporary, plywood flood-walls erected on top of the concrete jetties lining the river, but it was unclear whether these could hold. At last, the river crested just inches below the flood line, averting hundreds of millions of dollars in damages. Arial photos of the upstream regions south of town taken that day show thousands of acres of flooded forest and farmland: temporary wetlands, providing a quiet and profound demonstration of "ecosystem services." Had these natural sponges been paved over to make subdivisions, schools, and malls, there is little doubt that downtown Portland would have been the recipient of much of that water.

Other proposed decisions to convert natural habitat for development purposes in different parts of the world have been examined through a broad benefit-cost lens. Some recent examples: Analyses of forest lands in Malaysia and Cameroon and mangrove swamps in Thailand all found that the private benefits of conversion (to logging, agriculture, or aquaculture) exceeded the private benefits of preservation. However, when the social benefits of preservation were factored in (flood control, carbon storage, endangered species protection, shoreline stabilization), in each case preservation generated a larger flow of economic value than did conversion. Similarly, an analysis of wetland drainage in Canada for agriculture found that the practice was privately profitable (due in some measure to extensive subsidies), but that the lost value to society from hunting, fishing, and trapping exceeded these agricultural gains.

Ecosystem studies like these are complicated on the economic side by two very hard-to-measure values: existence value and option value. Existence value tries to incorporate human moral concern for species preservation into the dollar calculus of benefit-cost analysis. It does so by characterizing this concern as a form of "consumer demand." The idea is that people would be willing to pay to prevent the extinction of, say, a frog species—not because of any use value associated with the frog, but rather because they simply felt that the creature had a right to live. Existence value proxies the depth of this moral concern by an individual's willingness-to-pay for preservation.

On the face of it, this approach may seem inadequate, or even morally offensive. Doesn't monetizing morality lead us into the realm of pricing, and thus demeaning, the priceless? In defense of this practice, some economists argue that the biggest loss in welfare to humans from species extinction is often the sadness that humans experience associated with these events; conversely, the biggest motivation to pay for species preservation is moral concern. Thus, leaving existence value out of a benefit-cost analysis of ecosystem preservation means leaving out one of the biggest categories of benefit.

Option value is a second slippery but quite important value concept. Option value attempts to put a price tag on resources whose value is fundamentally uncertain. Suppose an area of land about to be opened up for grazing might contain some strands of wild maize with unique, disease-resistant genetic code. Or it might not. It makes sense for society to pay something (but not an unlimited amount) to preserve the land intact, holding on to the option that it does indeed harbor a valuable commodity.

The economic literature on "biodiversity prospecting" has tried to tackle this issue. What price should society be willing to pay to set aside land with unknown genetic resources? A handful of recent studies generates an extraordinarily broad range of answers for a single plot of land rich in biodiversity: from $20 to $9,000 for a hectare of rainforest in western Ecuador. The key factor determining this range is the fundamental uncertainty about the probability of getting a "hit"—a commercially valuable product. While existence and option value are important components of the flow of wealth provided by natural ecosystems, they are quite hard to measure, and as a consequence are not always included in economic studies of the benefits of natural diversity.

In spite of the complications in analyzing the value of natural wealth, a number of recent analyses have shown that the benefits (private plus social) of preserving intact ecosystems for their services and the biodiversity that they contain exceed the costs. And this observation holds more strongly if measures of existence and option value are included. Yet, even when scattered benefit-cost studies do argue for preservation, the reality remains that in most real-world cases, the benefits of preserving biodiversity flow to society broadly, and cannot, in reality, be captured by folks on the ground. This fact, in combination with damaging government subsidies, means that habitat conversion typically proceeds apace. To say that the private benefits of converting natural habitat to human uses exceeds the private costs is a truism. This fact (along with global heating) largely explains our potential trajectory through the mass extinction of the twenty-first century.

But even if economic analysis shows that, on the margin, the full costs of habitat loss (private plus social) exceed the benefits from conversion, this is a different thing than asserting that biodiversity loss will lead to human impoverishment across the globe. Even if many decisions to destroy habitat are, on balance, unsustainable, it is the sustainability of the entire economic system—measured by economists as the well-being of the typical individual—that matters. If rapid technological innovation can lower the costs of a broad array of goods and services, from food to health-care to manufactured products, then reductions in welfare from biodiversity loss might still be masked by a general march toward improved living standards.

Economic Catastrophe and the Web of Life

In his book *The Diversity of Life*, Harvard biologist E. O. Wilson asks this chapter's big question: "What difference does it make if some species are extinguished, if even half of all the species on earth disappear?" He points to the conventional use-values of species diversity (cures for diseases and pest outbreaks), as well as the economic importance of ecosystem services, and then directly takes on the logic of substitutability between natural and created capital:

> It is also possible for some to dream that people will go on living comfortably in a biologically impoverished world. They suppose that a prosthetic environment is within the power of technology, that human life can flourish in a completely humanized world, where medicines would all be synthesized from chemicals off the shelf, food grown from a few dozen domestic crop species, the atmosphere and climate regulated by computer-driven energy, and the earth made over until it becomes a literal spaceship rather than a metaphorical one.

Recall, however, that the question we are interested in is not human welfare in a fully humanized world, but one that is one-third humanized: a reduction of perhaps 35 percent of the planet's biodiversity. Some unique species-level resources clearly will be lost; dependent local ecosystems and the cultures based upon them will disappear. And yet, it is not obvious that a reduction in species diversity of this magnitude would leave insufficient numbers of critters and plants to support most species-level functions important for economic development: sources of food and industrial materials; genetic repositories for everyday agricultural, industrial, and medicinal needs. For the sake of argument, assume that for most uses, we can squeak by with the remaining millions of species.

This returns us to the web-of-life argument; at some point, a critical strand disappears and vital ecosystem services begin to collapse. Most of the scary potential scenarios deal with food production. The starting point for this discussion is the growth of global monoculture in agriculture. The green revolution that began in the 1950s had as its basis conventionally bred, high-yielding varieties of staple grains: wheat, rice, maize. By the 1990s, modern varieties accounted for more than 90 percent, 70 percent, and 60 percent respectively of the production of these crops in developing countries. In the United States, five broccoli hybrids account for more than 80 percent of the total acreage, with one hybrid species alone taking up more than half the acreage.

The fear is that relying on monocultures such as these makes food crops highly susceptible to diseases or pests. The Irish potato famine of the 1840s is illustrative; the nation relied on a single crop with a narrow genetic base. When a pathogen struck, it spread unimpeded, and over a million people starved to death. More recently, serious virus outbreaks in the global rice and maize stocks have been arrested by relying on resistant strains collected in seed banks. In the rice example, the strain that was resistant to the virus came from a single location; shortly after the species was collected and preserved, its former habitat was flooded by a hydroelectric dam. Now, along with increased reliance on monocultural crops, global heating is fostering the conditions for the rapid spread of pathogens and insect-borne diseases. Warming accelerates the breeding rates of insects, extends their geographic range, and increases the maturation rate of pathogens they carry.

The way to forestall this kind of catastrophe is to have on hand a sufficient stock of diverse crops to provide a suite of alternatives in the face of catastrophic diseases, pests, or climate-change–induced productivity drops. Crop diversity traditionally has been maintained in three ways: in the wild, on the farm, and in seed banks. The stock of wild repositories is hard to evaluate. Strains of wild soybean, tomatoes, coffee, and grapes are known to have gone extinct recently. With the growth of aquaculture, wild fish stocks—many of them also being depleted rapidly—have taken on new importance as a source of genetic material. Along with dwindling wild stocks of genetic diversity, on-farm diversity is also rapidly disappearing. In Bangladesh, the promotion of high-yielding varieties of rice has led to the disappearance of thousands of traditional rice varieties. In Mexico, farmers have lost about 80 percent of their traditional maize varieties over the last seventy years.

Seed banks are now thus the most common line of defense against crop disease, as well as the primary source for crop improvement. Some fourteen hundred banks around the world house about two million distinct seed varieties, with more than a third of this number stored at fifteen locations. But the banks are not all well maintained. Before it became a household word associated with prisoner abuse and torture, Abu Ghraib was a town that also housed Iraq's national seed bank, containing over fourteen hundred varieties. These were especially important given the country's unique historical role as the birthplace of agriculture. Many of the seeds were lost in the looting following the 2003 U.S. invasion, but in the mid-1990s, foresighted Iraqi scientists had sent two hundred specimens abroad for safekeeping, and there is hope that others eventually can be replaced. These seeds will form the foundation of agricultural recovery in the war-torn land.

The Iraqi case is extreme, but in general, developing-country seed banks are facing difficulties simply maintaining the varieties already under their care. Even seed banks in developed countries face challenges. Over the last ten years, only one-third of the banks in wealthy countries have seen budget increases, although more than 80 percent have increased the size of their collections.

Given this background, consider how agricultural scientists likely would respond to the emergence of a serious pest or virus: look first to seed banks, second to known strains on farm, and finally— and likely when it was too late—send teams off bioprospecting for unknown (by definition) wild varieties. Put another way, if the wild strains were not already catalogued in gene banks, it is unlikely that they would be discovered in time to head off an imminent viral plague. This logic implies that the threat of catastrophic crop failure arises fundamentally from the adoption of monoculture, and only indirectly from loss of wild diversity. As an insurance policy against this kind of disaster, habitat likely to contain wild food strains should be explored thoroughly, and samples stored in (well-funded) gene banks, prior to any habitat conversion. Of course, maintaining habitat intact provides a more secure back-up storage system should seed banks fail.

Monoculture has its benefits: Standardized production methods allow agribusinesses to mass-produce farm supplies, and cookie-cutter production techniques lower management costs for farmers. Against these private benefits, however, must be weighed social costs, and one of these is a potential vulnerability to wide-scale—regional or even global—infestations of pests, diseases, or invasive species. Natural and on-farm diversity in the past has provided firewalls against a catastrophic food-system failure, but, increasingly, fewer and fewer strains are dominating global agriculture. At the same time, in the face of human-induced climate change, the spread of diseases and pests has begun to accelerate. Ominously, a recent report by the Council for Agricultural Science and Technology noted the inherent risk of a monoculture strategy: "Even the full complement of natural genetic variation may not be sufficient to stop some diseases."

How likely is a monoculture disaster on the scale of the Irish potato famine? No one knows. Ecologists themselves debate the resilience of ecosystem functions. Elizabeth Willot, an ecologist and expert on mosquitoes, was asked by a reporter the big question about the ecological significance of her particular creatures: "Would everything really collapse if we got rid of them? Well, she said, no. The web of life is not that fragile. 'If you take a snip, it won't unravel.' In fact, she said, there is 'quite a bit of ecological research now showing that removal of a species doesn't make a huge difference.' If the species of mosquitoes that are intimately connected with human beings were made to disappear, there might be some ecological disturbance, but 'you probably could remove them without catastrophe.'"

E. O. Wilson concurs on the narrow point: "The loss of the ivory-billed woodpecker has had no discernible effect on American prosperity. A rare flower or moss could vanish from the Catskill forest without diminishing the region's filtration capacity." But for Wilson, these facts miss the point. "The following rule is now widely accepted by ecologists: the more species that inhabit an ecosystem, such as a forest or lake, the more productive and stable is the ecosystem." And Wilson warns that continuing to compromise this productivity "puts at risk not just the biosphere, but humanity itself."

Regardless of the resolution of the scientific debate on the fragility, or otherwise, of the web of life, the economic case for habitat preservation is straightforward. As we saw in the previous section, defending the diversity of life often makes economic sense from a simple analysis of marginal benefits and costs to society. On an increasingly full planet, habitat conversion can compromise day-to-day ecosystem services in ways that leave us worse off, even barring catastrophe. In this context, the somewhat uncertain threat of a serious failure of the food system, or of other ecosystem services, is an additional economic argument for habitat preservation—as well as for active collection programs and the maintenance of gene banks. The sensible economic response to the threat of catastrophic events is to buy insurance, and the protection of genetic diversity provides some measure of that.

The Economic Meaning of Life

Among people who study global heating, and among those who fight for solutions, there is often a temptation to give in to despair and hopelessness. This crisis seems to demand unprecedented vision and foresight on the part of humanity and their governments. Apocalyptic visions of the future have been the springboard of environmental awareness since the 1960s, a decade bookended by Rachel Carson's *Silent Spring* and Paul Ehrlich's *Population Bomb*.

But because our past environmental problems all have been at least perceived to be local, solutions always seemed to be within our grasp. Thanks to a deliberate human response, in part provoked by those visions, Carson's nightmare never came to pass; Ehrlich's bomb has been, if not defused, contained.

Global heating is different. The human response it calls for is truly heroic, requiring nothing short of rewiring the entire planet with a new generation of clean-energy technologies—and doing that very soon. The task seems too big; our social and political tools too clumsy. Are we, as a species, capable of this kind of deliberate global response? With Keynes, I remain a qualified optimist. Capitalism, in a popularly responsive political context, remains an incredibly resilient and dynamic system, with a relentless underlying drive toward the accumulation of capital. This capital, in turn, produces a growing flow of goods and services to feed, house, clothe, educate, and care for a world population that, with effort and luck, will stabilize by the end of the century at a manageable ten billion people. Capitalism, properly harnessed, can also deliver and diffuse the clean-energy technologies that we need to stabilize the climate. Even more so than did Keynes, I believe that it will be politics that determines the material welfare of future generations: the pace at which we advance toward "our destination of economic bliss," and perhaps, whether we continue to advance at all.

Oil stocks and especially climate stability are natural resources for which significant shortages are looming. If unaddressed by the world's political systems, these constraints could generate serious economic setbacks, accompanied by destabilizing political turmoil. Loss of biodiversity on its own, in my judgment, seems to pose a less-dramatic threat. There is some risk of catastrophic impacts working through the food system, but more likely is a gradual erosion in well-being as a broad array of ecosystem services is further frayed by conversion of natural habitat and the mounting impacts of global heating.

Having said that, the preservation of diversity increasingly is understood as making good economic sense. A recent study by a team of ecologists and economists looked at the trade-offs between farm and forest income and the preservation of close to a hundred terrestrial species inhabiting the working landscape of the Willamette River Valley, just south of my home town of Portland, Oregon. They found that "intelligent land use planning"—planning that takes into account the range needs and dispersal patterns of the animals—can provide suitable habitat for the vast majority of species and still maintain rural incomes at high levels. They concluded:

> Even the limited trade-offs between conservation and economic objectives shown in the example above may be something of an overstatement. In this paper we did not consider the economic value of ecosystem services, such as the provision of clean water, nutrient filtration, climate regulation and ecotourism. Including the value of ecosystem services . . . would tend to reduce apparent trade-offs between conservation and economic objectives further.

Life has meaning to humans most immediately as a direct and generous provider of wealth. The key to protecting that wealth is, again, smart political decisions ("intelligent land use planning"). And, if loss of genetic diversity seems less likely than other environmental constraints to impoverish us dramatically, it nevertheless will have large economic impacts. Each unique habitat we lose closes off options for our grandchildren. Preserving diversity is thus best thought of less as a way to head off an imminent economic collapse, and more as a legacy—a gift of knowledge that might bring wealth we cannot imagine to our descendants.

CHAPTER 3

KNOWLEDGE

On the south side of the mountain I grew up on, about five miles from the Alabama border, there is a roadcut that teems with the remnants of life. If you sit on the sandstone shelf and run your fingers through the debris that has weathered off the hillside, you find fragments of a 320-million-year-old ocean floor: the stems of sea flowers; coral colonies forming small, perfect screws; and little round nut-like creatures, distant kin of starfish. Most of these fossils have recognizable living relatives, but the remains of one creature are quite strange to modern eyes. Like an armored, sea-faring beetle, the trilobite etched in its stone casting is mute testimony to the meaning of extinction. There is nothing like it alive today on the planet.

If you scramble a short way uphill from the roadcut to the ridge, and from there find a cool, heavily shaded limestone-lined creekbed, you can follow that down a mile or so to Lost Cove, an unlikely valley hidden halfway up the mountain. In the heart of the cove is a stunning cave opening, with layered limestone and sandstone cliffs rising a hundred feet above the wide creek that flows from the cathedral-like entrance out of the cavern. Native Americans inhabited this place for thousands of years; you can find shards of pottery and arrowheads under the overhangs.

Here are some of the things I learn from in that cove. Oak and poplar, maple and hickory, cedar and hemlock and sycamore and locust and dogwood. Thick wild grape vines and fields of poison oak; clusters of black-eyed susans, and purple and white and pink three-leaved trillium. Bees. Mountain laurel and huckleberry bushes bearing tiny black bitter fruit. Deer and squirrels and raccoons and possum and porcupine. Two copperhead snakes, stretched out, entwined , making love. Toads and turtles and snails. In perfect webs, nameless spiders. In the cave, sleeping folded bats, albino crawdads, and albino crickets. In the leaf litter, dozens of tiny scorpions, solitary beetles, centipedes. Tree bark covered with busy ants; tree stumps loaded with fat termites. Night crawlers and leaches and earthworms. Gartersnakes and rattlesnakes and black racers. Pileated woodpeckers and bluejays, cardinals, hawks and crows and buzzards and owls, and innumerable, anonymous little grey birds.

Lost Cove is where I go to school. Some of the knowledge that I gain there has practical value: I know which snakes to play with and which to admire from a distance. But much of what I learn doesn't seem to yield an immediate payback beyond personal satisfaction: identifying a hickory by its shaggy bark or a sycamore by its spiny fruit, what do those particular lives mean to me?

Gecko Feet, Sewage, and Hope for the Future

One serious student of nature works in the building next to mine. There, biologist Kellar Autumn and his research team recently unlocked the secret of gecko glue. Geckos have the ability to walk up walls and even to hang upside down on polished glass. Autumn learned how this is possible: Each gecko foot is loaded with half a million tiny hairs called setae, each the length of the diameter of two human hairs. Setae are incredibly sticky. Geckos can support their weight comfortably by one toe. If a gecko had enough muscle power, the adhesive grip in his four feet—a surface area the size of a dime—would allow him to lift a small child clean off the ground. So how does it all work?

Geckos apparently tap an intermolecular attraction called the van der Waals force. At the end of each setae are hundreds of thousands of pads called spatulae. As the spatuale come in contact with a surface, unbalanced electrical charges around the molecules in the gecko's foot attract similarly unbalanced charges in the wall, drawing the two together. By rolling the pads on its feet, the gecko can stick to and then peel itself off perfectly smooth surfaces. Autumn's analogy: "It's kind of like geckos have Velcro, but without the other side."

The mention of Velcro takes us quickly toward practical application. And the potential uses of gecko glue are certainly intriguing. One project on the drawing boards is a tape that sticks in the vacuum of outer space. Autumn's team is also partnering with a robotics firm to create geckobots that could be used in search-and-rescue operations or as children's toys. Synthetic setae might be used for micro-surgery. And then there are fumble-proof gloves, and of course, climbing gear. "I think we'll be able to do better than 'Spiderman' someday," said Autumn.

Worldwide, there are at least a thousand species of geckos. In its most recent "red-list" assessment, the International Union for Conservation of Nature and Natural Species examined twenty gecko species thought to be vulnerable to extinction. (This is not a comprehensive survey, just an evaluation of species that had been studied.) Of these twenty, one was extinct, another critically endangered, and only three species were deemed to be relatively healthy. The gecko family itself is not in danger of disappearing from the planet anytime soon. As individual species go extinct, however, opportunities for learning about subtle or perhaps dramatic variations in the glue fade as well. And over the next few decades, as more and more lizards are threatened, the pace of those lost opportunities will accelerate.

Suppose geckos had in fact become extinct three years ago, before Autumn's discovery. The extinction would have made no more than a ripple in the daily news, but Autumn notes, "An adhesive nanostructure is such a bizarre concept that if geckos hadn't evolved it, it's likely that humans would never have thought to invent it." And if gecko glue was never to be discovered, then humans would have been poorer—but not impoverished. Humans would have prospered even without the invention of Velcro. Yet much more so than Velcro, the principle behind gecko glue may, over the next century, underlie engineering applications that become as ubiquitous as cell phones.

The natural world is a treasure trove of knowledge, some of it with obvious potential for commercial application, but much of it still beyond our ability to analyze or comprehend. Through trial and error, natural selection has been patiently solving baffling engineering and medical problems for four billion years. The fruits of nature's efforts are stored in the genetic code of individual species, embedded in the living library of creation. One hundred years ago, Theodore Roosevelt's comment on extinction anticipated a similar understanding of loss: "When I hear of the destruction of species, I feel just as if all the works of some great writer had perished." Autumn expects that nature holds many more adhesion secrets that can be put to work. "Biodiversity is like this library of novel engineering applications, but extinction is taking these books off the shelf faster than we can read them. These are valuable secrets of nature that are being lost forever."

It is not only from individual creatures like geckos that we can learn. Interacting, complex natural systems too provide important lessons. Take sewage treatment, for example. Today, lack of adequate sanitation and safe drinking water is the number one environmental health problem in the developing world: Each year, some 1.7 million people die from bad water, with children accounting for 90 percent of the fatalities. Over the next fifty years, as the human population expands to (and, one hopes, stabilizes at) ten billion souls, a major environmental challenge will be dealing with all that extra human waste. And not just human: Feeding those additional mouths will mean a dramatic increase in factory farming of chicken, poultry, cattle, and fish. These operations are already generating huge waste problems of their own. In the United States alone, farm animals currently produce over two trillion pounds of waste annually. The seven million factory-raised hogs living in North Carolina produce four times as much sewage as the state's 6.5 million people.

The conventional approach to human sewage treatment is to disinfect wastewater using chlorine. While effective in controlling disease among humans, chlorine can also corrode surfaces and is toxic to fish, while chlorine gas can be a serious respiratory hazard for treatment plant workers. Animal waste is treated with much less care. Generally, the waste is pumped into collecting ponds, and then liquid manure is sprayed onto fields. The quantity of waste often is more than the crops can utilize, and the excess runs off into surface water or escapes into the air. "Nonpoint source" pollution, much of it from agriculture, is the primary source of water contamination in the United States today. In a world where sewage volumes are likely to double at the same time that surface water demands are increasing, better methods of protecting water quality from sewage pollution clearly are needed.

Nature provides a model. In modest amounts, human and animal waste is digested easily by the natural world, and indeed, is food for the ecosystem. Human engineers have been exploring nature's digestive capabilities, and recently have designed high-volume natural sewage treatment "plants," or rather, carefully orchestrated collections of plants and animals. Dubbed "Living Machines" by inventor John Todd, these engineered wetlands produce fresh water without the use of chemical additives.

Garfield the Cat facilitated an early test of one of these systems. The comic strip character is supported by a company called Paws Inc., employing fifty people in the marketing of his feline-related paraphernalia. When the company relocated to rural Indiana in the late 1980s, no sewage system was in place to handle the waste volume from the new company headquarters. So Garfield's creator, Jim Davis, had a solar aquatic system designed and built. The engineers combined various forms of bacteria, algae, water hyacinths, duckweed, floating ferns, snails, Japanese blood grass, papyrus, hibiscus, watercress, caladium, Angel's trumpet, and a mélange of other temperate and tropical plants in a series of carefully designed treatment pools. Four species of fish prowl the final lagoon, where hybrid tea roses are harvested to decorate the company offices with fresh cut flowers.

The U.S. Environmental Protection Agency has analyzed several demonstration projects for Living Machines and has found that they are largely effective in cleaning up wastewater without relying on chemical inputs, and that they can provide cost savings for some communities, even at this stage of development. In warm climates, Living Machines are cost competitive up to flows of about 1 million gallons per day, serving a city of about ten thousand people; in temperate climates, where a greenhouse must be constructed to protect the ecosystem, solar aquatic systems can compete with conventional treatment facilities for flows of 600,000 gallons per day. And at the end of the rather dry EPA cost analyses, one finds the following observation: "Its unique and aesthetically pleasing appearance make it an ideal system in areas that oppose traditional treatment operations based on aesthetics (smell and appearance)."

It is perhaps odd to be inspired by sewage, but in fact, the general principle behind the Living Machine provides the key to envisioning a sustainable future for humankind. For people who pay attention to the natural world, hope is a critical commodity; it is often difficult to sustain against the relentless pressures of population and consumption growth. In his book *The Ecology of Commerce,* businessman Paul Hawken guides a route through the next century, explaining clearly how humans, now six and soon to be ten billion of us, can live sustainably on the planet. Quite simply, businesses must transform themselves so that their operations mimic nature. In sustainable commercial systems, as in nature:

1. all waste is food;
2. energy is derived from "solar income" not "solar wealth";
3. a healthy diversity of entities are supported.

The first principle is a mandate for "closed-loop," zero-pollution production, in which the waste or effluent from one production process becomes the input into another production process. The leaf litter on the forest floor in Lost Cove is food for insects; their digestive processes in turn produces "waste" that enriches the soil. A solar aquatic sewage system is, in effect, a complex machine for converting waste to food, to waste, to food, to waste, to food, throughout the web of a carefully designed ecosystem. Unlike conventional water treatment processes that discharge chlorinated water, nothing in the final product of living machines is not food for another organism.

"Solar income" refers to the direct power produced every day from the Sun. This means electricity produced by solar cells and wind mills (because energy from the Sun creates wind), and also solar energy trapped by plants and converted to biofuels such as biodiesel, ethanol, or wood. Solar wealth, by contrast, means energy from the Sun that has been trapped over the eons under the earth in deposits of oil, coal, and natural gas. This energy is like wealth in a bank account left to us by a prosperous ancestor—but it is one that we are fast depleting. The forests and animal life in Lost Cove do not run on fossil fuels; instead, they depend on the daily, direct input of the Sun. Living Machines also run largely on solar power, as most of the clean-up work is done by plants directly converting sunlight to energy via photosynthesis.

Finally, Hawken observes that healthy natural ecosystems support a diversity of life forms. In contrast, the monocultural entities of modern global business, the Wal-Marts and McDonald's, Hawken feels are unnatural creations, succeeding only through a variety of hidden subsidies. In a sustainable commercial system, as in a healthy ecosystem, a diversity of business forms would thrive.

Hawken's vison of our current, rapacious system of business enterprise transformed into an "ecology of commerce" is a powerful, inspiring metaphor. And his three commercial ideals, the result of his own close observation of the natural world, provide both the best available road map toward sustainability, and a guide for smart government policy. Over the next half century, combined increases in population and affluence might lead to a quadrupling of the human footprint on the planet. In the face of this growth, technological innovation that pushes us in these principled directions will be critical to protect and restore natural ecosystems. But such innovations do not come "naturally." Ultimately, government will need to rewrite the rules of the game to help the commercial system mimic nature.

So where do we get when we start differentiating between hickory trees and sycamores? Spiderman glue, clean water for comic cats, and a plan to save the planet. This is knowledge produced from an intimate familiarity with the natural world that ultimately may produce material wealth: convenient gadgets, aesthetic surroundings, longer, healthier lives. But it is also knowledge that is fascinating and uplifting on its own terms. Kellar Autumn's research team is aware of the commercial applicability of their work, but is driven by a deep desire to observe the unobservable, to unlock the engineering secrets that nature has developed so patiently. John Todd and his successors are inspired less by the money to be made from Living Machines than by building a truly beautiful synthesis of natural and commercial systems. And the only tangible thing that I have gained from Paul Hawken's meta-theorizing about the ecology of commerce is a glimpse of a practical way through the daunting environmental dilemma facing our children's generation. The small thing I have gained is, simply, hope for the future.

But what does all this have to do with the sixth great mass extinction? When and if humans stabilize both our population and our rampant destruction of other species on the planet, won't there still be enough of nature's creation around to serve as a platform for innovation and aesthetic endeavor, and to satisfy the human thirst for knowledge? Certainly, our descendants will work whatever ground we leave them, no matter how impoverished. But every lost species leaves a permanently smaller platform, a diminished palette not just for our children and grandchildren, but for the countless generations of artists and engineers who will follow them, forever, until the end of history.

Megalosses

"Charismatic megafauna" is an invented bit of jargon, originally sort of an inside joke. First employed by professional environmentalists (who go by their own special nickname of "enviros"), the term refers to large, often cute or awe-inspiring, often furry or feathered creatures who are the poster children of endangered ecosystems: spotted owls, polar and panda bears, wild elephants, chimpanzees, or African lions. Sometimes, charismatic megafauna can even have scales or shells: think Northwest salmon or Hawaiian sea turtles. Endangered *Abrawayaomys ruschii* (Rushi's rats) are definitely neither charismatic nor megafauna.

The term persists, I think, because the word "charisma" is so fitting. I have spent a little time close with animals in the wild. On one glorious trip in southeast Alaska, we were winded-in on a rocky shoreline with a monstrous black bear stalking the alder thickets on the upper beach; and we shared a dead-calm fjord with a bald eagle perched high upon an iceberg; and most thrilling, we had our kayak shadowed by a pair of humpbacked whales, and then watched as they came closer, and then dove directly under the thin fabric of our boat, their massive striped bodies streaking just a few feet below us, finally flashing magnificent tail flukes a couple of dozen yards off the starboard. What proximity to all these animals instills, as with charismatic humans, is immediate, deep respect, a compelling desire to remain in their presence, and to learn more from them.

To the traditional inhabitants of Alaska, charismatic megafauna also held and hold a special appeal. The anthropologist Richard Nelson lived with the Koyukon people in interior Northern Alaska during the mid 1970s, when most of the locals over thirty still spoke the native language and when a subsistence economy— albeit modified by snowmobiles and cash infusions from migrant workers—still dominated the culture. Nelson learned about the complex natural history of the northern boreal forest, the exquisite ecological knowledge that undergirds the Koyukon economy, and the cultural and spiritual worldview that such a lifestyle engenders.

The Koyukon closely observed their local megafauna—moose, otter, fox, owl, wolf, lemming, shrew, black bear, brown bear, ptarmigan, whitefish, weasel, marten, marmot, salmon, hare, raven, wolverine—and told complex stories about their origins and interrelationships. These observations in turn shaped their own self-understanding. The Koyukon taxonomic system, for example, judged relatedness of species based on behavioral characteristics, rather than on morphology. For example, the black bear and the porcupine, who occasionally will share a winter den, were considered closely related based on their ability to "get along." Dogs and wolves, however, with very different temperaments, were not thought to be near-relatives. Indeed, dogs and *humans* were quite closely related. According to Koyukon stories, in "distant time," dogs and humans could speak to one another. Dog owners were referred to as the grandfathers and grandmothers of their working animals, and the Koyukon ascribed powers to dogs such as the ability to foretell death.

In Western society, animal behaviorists, ecologists, conservation biologists, wildlife photographers, and people who fish, hunt, and hike engage in similar processes of observation and storytelling, learning from contact with wild animals about themselves and about the human species. In a wonderful book called *Sex Advice for All Creatures,* the author, "Dr Tatiana," collects some of the tales told by modern evolutionary biologists. Many of the most compelling stories are about our genetically nearest neighbors: chimps and bonobos.

Most people have not even heard of a bonobo—these apes were only discovered in the 1930s, and were not studied in the wild until the 1970s. And yet a recent debate has raged among anthropologists as to which of the two species are more like our common ancestor. The behavioral evidence is intriguing: Like humans, but unlike bonobos, chimps are tool-makers, participate in group hunting, engage in lethal raids on other chimp bands, and exhibit both intense male bonding and male dominance over females. And like humans, but unlike chimps, bonobos enjoy recreational, nonprocreative sex, have strong friendships among females, exhibit relatively egalitarian male relationships, and are capable of "relaxed intergroup interactions." The relationships among chimps, humans, and bonobos are rife with cultural puzzles that early field observations are only beginning to untangle.

On the sexual front specifically, with the advent of DNA testing and a few decades of observation, we now know that chimpanzee women are quite promiscuous and that female bonobos are generally bisexual. What is the meaning of these observations for primate sexuality as an evolutionary strategy, and ultimately, for human sexual self-understanding? At this point, no one really knows. The data are too new to have yet shaped our collective sense of self. But these new stories about our nearest genetic relatives clearly take us beyond our old cultural boundary, framed by the pairing in Genesis of creatures two by two.

Surprisingly, it was neither a chimp nor a bonobo but a gorilla with whom Westerners had their first interspecies conversation. Since 1972, the mountain gorilla Koko has been talking with her scientist pals in sign language (her partner in conversation, Michael, died in 2000). Koko, who has a vocabulary of about fifteen hundred words, scores between 85 and 95 on IQ tests; the average human score is about 100. Frances Patterson, who has worked closely with the gorillas, summarizes her experience this way:

> Koko and Michael have created representational art in which they depict emotions, events (such as an earthquake), or cherished friends or pets (Koko painted her bird, Michael painted his dog), and they use language to name those pictures. They use and construct tools. They make-believe, as when Koko acts out fighting or biting scenes with her dogs or alligators, and or pretends to nurse a monkey doll or make her ape doll sign. Sometimes we are asked to play along with these scenarios, "drinking" from empty cups, or reacting as if bitten when Koko threatens us with a toy alligator. She seemingly expresses grief, pride, shame, embarrassment, humor and deception. She appears to attribute mental states to others. Collectively these features create a strong case for gorilla personhood.

The Koyukon believe that they could once talk to dogs; we have, in fact, recently discovered how to talk to gorillas. If we can keep them from dying out, and carefully learn enough about them, perhaps we can someday talk with bonobos. And then, perhaps, with whales or dolphins. And in the process, slowly, as the Koyukon did, we might expand the circle of persons and near-persons.

The "knowledge" reasons for preserving megafauna are the same as for preserving species generally. We have much to learn from their social interactions, their physiological blueprints, and their genetic code—all products of hundreds of millions of years of evolution, all reflecting nature's finely honed and remarkably engineered solutions to complex survival challenges. But the charismatic creatures that are closest to us genetically provide truly special opportunities. Recent research, for example, has focused on older apes. In the United States, there are about forty-five captive apes older than forty. This tiny collection of chimps, gorillas, bonobos, and orangutans is providing insight into the development of diabetes, obesity, and Parkinson's and Alzheimer's diseases. As populations and species of the most highly evolved mammals—cetaceans, apes, bears, big cats, wolves—die back, we close off whole avenues for knowledge that are simply irreplaceable.

Megafauna are at special risk because of their size: They require large tracts of unbroken, often specialized habitat. As forests, savannah, and tundra are logged, brought into cultivation, converted to mining and oil and gas development, or are dissected by roads, habitat is destroyed and fragmented. Megafauna also face pressures from overhunting: The bushmeat trade threatens the existence of gorillas, chimps, and bonobos. About three to six thousand gorillas are killed by poachers each year, decimating the few remaining wild populations. Black bears, big cats, and wolves have a difficult time co-existing with suburbanites and ranchers.

Even for those creatures who inhabit the still-sparsely populated region above the Arctic circle, global heating now looms large as a threat to their existence. As noted in the introduction to this book, researchers have been documenting the slow starvation of the polar bear along Canada's Hudson Bay. As the ice pack breaks up earlier each year, the hunting ground for the animals is disappearing. When mothers become too lean, they stop nursing, and their cubs die.

For the Koyukons—as for Native Americans broadly—bears are the creatures with the closest claim to personhood in the forest. Mother bears are fiercely maternal, and also like humans, cubs pass through an extended childhood, suggesting a vast capacity for learned behavior. Bears share a quite similar diet with humans. Roughly 70 percent vegetarian, they are master gatherers. Bears can walk on two feet, leaving prints that mirror those of people, and—as virtually every bear observer notes—a skinned bear looks disturbingly like a human corpse. For all these reasons, bears are a species that share "something akin to humanity, some special behavior and consciousness that in the Koyukon perception moves beyond what is purely animal."

For good evolutionary reasons, the charismatic creatures with whom we share the planet compel us to focus on them, and to study their ways. And now, with a whole new suite of scientific tools at our disposal, we discover that we have barely begun to unlock their secrets. A Koyukon man told Richard Nelson: "Each animal knows way more than we do. We always heard that from the old people who told us never to bother anything unless we really needed it."

Ways of Knowing

As wild creatures and their habitat disappear, opportunities for knowledge also vanish. But, just as tragically, so do different ways of knowing. The traditional human cultures that co-evolved with expansive and stable ecosystems on all six continents—environments all under dramatic siege the last two centuries—now also face extinction in the twenty-first. Hunter-gatherer cultures are the exemplars of the original human social form, one that defined 99 percent of our evolutionary history.

Belatedly, Westerners have begun to recognize that traditional people have accumulated detailed and imminently useful biological and cultural knowledge: Shamans in the upper Amazon basin have become used to regular visits from pharmaceutical representatives and bioprospectors. The loss of these communities will mean the loss of this ecological knowledge, and the disappearance as well of the anthropologist's primary laboratory for gleaning insight into the universals of human behavior along dimensions of kinship and sexuality, altruism and spiritual practice. But subsistence societies offer more than just medical and cultural knowledge. Embedded in the intricacies of their language and ritual, they carry forward different ways of knowing.

As global culture homogenizes into the twin forms that Benjamin Barber has called Jihad and McWorld (tribalistic theocracy and globalized secular capitalism), the oldest human worldview, one that was expressed in diverse forms across the planet, and that featured, everywhere, an enforced and intimate relation with nature, is rapidly disappearing. Richard Nelson reflected on what he had learned from his years immersed in the subsistence culture of Alaska's interior: "As I was living among the Koyukon people, nothing struck me more forcefully than the fact that they *experience* a different reality in the natural world . . . For the Koyukon, there is a different existence in the forest, something fully actualized within their physical and emotional senses, yet entirely beyond those of outsiders." He goes on to say:

> From this perspective, much of the human lifeway over the past several million years lies beyond the grasp of urbanized Western peoples. And if we hope to understand what is fundamental in that lifeway, we must look to traditions far different than our own . . . Probably no society has been so deeply alienated as ours from the community of nature, has viewed the natural world from a greater distance of mind, has lapsed to a murkier comprehension of its connections with the sustaining environment. Because of this, we are greatly disadvantaged in our efforts to understand the basic human affinity for nonhuman life. Here again I believe it is essential that we learn from traditional societies, especially those in which most people experience daily and intimate contact with the land.

It is easy to romanticize hunter-gatherer and subsistence cultures. But it is also easy to dismiss as superficial what are fundamental differences in ways of understanding the world. It appears that hunter-gatherer cultures do cultivate in a profound way the human connection to nature, and often develop a fundamentally biocentric worldview, one that interweaves human and natural worlds into the same tapestry. Nelson notes, "The Koyukons seem to conceptualize humans and animals as very similar beings. This derives not so much from the animal nature of humans as from the human nature of animals."

How transferable to Western culture is this kind of deep connection to nature? The philosopher Bryan Norton suggests one view in his assessment of the potential transformative power of the natural world:

> A better understanding of the true human role in ecosystems would encourage belief in a more rational world view, one that clearly recognizes that the human species as it now exists is an evolutionary product of natural, environmental forces and is dependent on the survival of other species for its own survival. Encounters with wild species and natural ecosystems encourage acceptance of [this] ecological world view and cause humans to question the value of unlimited consumption.

Norton's argument is that experience in nature can produce a rational understanding of ecological interconnectedness and constraints, and a consequent recognition that humans are damaging natural ecosystems beyond repair. Adopting this ecological worldview will in turn lead to rejection of "overly materialistic and consumptive felt preferences." He suggests that as a consequence, we individually might make a commitment to renounce our high-consumption lifestyles. This would be Thoreau at Walden writ large: The pond taught him the true nature of economics, which was to simplify one's life in order to enrich it.

Based on a survey of my environmentalist friends, this process unfortunately seldom unfolds as described. While many of us nature lovers drive hybrid cars and change our incandescent light bulbs for compact fluorescent ones, we also remain affluent Americans. We haul our kids from our spacious suburban houses to soccer games and orthodontist appointments. And we spend a lot of our money on travel. Two round-trip, cross-country plane tickets wipe out the annual benefits of the hybrid car, giving us as large an environmental footprint as many of our less well-traveled, SUV-driving neighbors.

Norton's hypothesis fails for two reasons. First, as I suggested in the previous chapter, even to those with ecological knowledge, it is not obvious that we humans are dependent on the survival of all or perhaps even most other species for our own survival. Thus the moral imperative to protect species is not as self-evidently self-interested as Norton suggests. Second, it is also not clear that experience in nature leads to an essentially different *rational* understanding of ecological systems. In an interesting study of children's attitudes toward the environment, Peter Kahn looked for cross-cultural variations in both depth of concern and type of ecological reasoning (utilitarian versus biocentric). His team found no significant differences among kids raised in inner city Houston, the sprawling Brazilian city of Manaus, and the small, isolated town of Novo Aryao, eight hours up-river on the Amazon from Manaus, and accessible only by boat. In many cases, the children's responses from the different communities closely echoed one another.

Kahn offered one explanation for the lack of contrast: "While the village was accessible only by boat, it was still heavily influenced by the missionary culture." In other words, the kids in even the remote Amazon essentially had been Westernized. He suggests that the results might have differed if he had worked with an indigenous hunter-gatherer community like the Koyukon. Nevertheless, Kahn points out that "the children in Novo Aryao lived close to the land, with daily and intimate contact, and that by itself was not enough apparently to engender a biocentric reasoning." The conclusion is that real, felt biocentrism will only emerge "naturally" through a combination of sustained immersion in nature *and* a supportive culture.

Norton argues that contact with nature encourages a more rational worldview: As we become natural ecologists, recognizing deep links and fundamental limits, we are then forced to rethink and change our high-consumption lifestyles. I agree with Norton that nature has transformative power, but believe that the power is not one of higher rationality. Instead, what subsistence cultures can teach about is what Nelson identifies as lying at the heart of Koyukon identity: "the basic human affinity for nonhuman life." Westerners who hike, hunt, fish, or camp on weekends, or who work on farms and in the woods, get a taste of that affinity. Hunter-gatherer cultures, fast disappearing, can provide humanity with a diverse set of menus for the full-blown banquet.

I agree with Norton as well that experience in nature can be an essential key to changing human behavior. But I believe that those behavioral changes arise less from a rational recognition of self-interest in species preservation than from a mysterious falling in love with the natural world. This, however, is where knowledge blends into ways of knowing and then into spirituality—which is the subject of the next chapter.

Traditional knowledge is found not just in indigenous communities. From fathers, scoutmasters, old-timers, and older boys who had learned from older boys, I know that green hickory branches are supple and will bend without breaking; that you can peel cedar bark for good fire tinder on wet days; how to gut fish; that poisonous snakes have distinctly triangular heads; and where and when, in the mountains, you are likely to find a spring, ripe berries, or flat, dry ground. Of course, from northern Alaska to south-central Tennessee to the Amazon basin, as traditional rural economies disintegrate, the old-timers are disappearing too.

Today, however, there is a new and overarching threat to traditional knowledge. As the climate changes, the forests and the mountains and the meadows are themselves changing, slowly now, but at a pace that will accelerate dramatically over the century. Global heating potentially threatens changes in temperature and habitat on the scale of an ice age over the next hundred years— though creating a hotter and not a colder home. The very existence of traditional knowledge, in short, assumes a stability in the natural world that may, like species and cultures, simply slip away.

A year ago, a friend and I spent a couple of days exploring the north flank of Mount Adams in southern Washington. Two miles above our camp, the Adams glacier rolled off the summit, falling in inches-to-the-year slow motion, a series of cascading icefields and icefalls that hung silently in the air, defying gravity. Melt from the glacier fed a ferocious stream by our camp, thick and muddy with brown silt. Since it was the height of the dry season, we hadn't brought any rain gear. (It is a well-kept Northwest secret that the months from July to October feature largely unbroken sunshine and cloudless skies). But in the morning, mist rolled in, and by nightfall a steady rain had driven us into the tent. At daylight the next day, with the rain still falling, we packed our gear and hiked back out to the truck. The stream, swollen by the accelerated snowmelt from the warm rain, had doubled in width.

As we headed down the road, a parked van gave us a honk. A man appeared at the window. His vehicle had busted an axle on the rough road, and the man, his grandniece, and a buddy had spent the night in the van. His name was Ray; he had come up to the mountain to join his son's family on a berry-picking expedition.

Ray was about fifty-five, and while we drove for an hour down to Summit Lakes to find his family, he told us a little bit about himself. He was a Warm Springs tribal member. As a young boy, he recalled his family being relocated when the Army Corps put in the last of the dams on the lower Snake River. From his mother, Ray had gained a lot of traditional knowledge. He spoke the Indian language fluently and was thinking about heading back to community college to pick up the certificate he would need to teach. In the summer, he followed a semisubsistence lifestyle—fishing and drying salmon, digging camas roots and wild potatoes, and spending late August and September picking huckleberries. Ray had passed the hours in the van before we showed up on a side business, making baskets out of cedar bark

We took Ray to a camp on the edge of what is now officially known as "Indian Heaven Wilderness." For centuries, native people have come here from all over the Northwest in August and September to harvest huckleberries and hold summer celebrations. At one spot, you can see the remains of an oval racetrack, beaten into the ground under the annual thunder of horses' hooves. Ray had been coming here for decades, spending weeks or a month at the same campsite. When we arrived, still in the rain, a couple of dozen families were scattered across the camps. Ray found some friends to help him with the van. People were sleepy. There had been an all-night Native American Church service. We ate a couple of donuts and headed back to Portland.

Against the onslaught of television, fast food, and video games, some American Indians work to bring up their children into their heritage. Harvesting huckleberry, salmon, and camas give heart to that cultural tradition. Can anyone doubt the value to a young person of weeks spent with family, playing and working out of doors, in a place that has welcomed their parents and their parents' parents before them? But these life-sustaining foundations of creation, woven into the fabric of ten thousand years of Northwest culture, are disappearing. Ray noted that years of fire suppression have compressed the range of the berries; as we drove, he pointed out alder-covered slopes that he remembers as once being rich in fruit. At the end of the season, Indians used to set fire to the slopes. Some men, Ray said, knew how to call the rain to control the burns. My Western interpretation of that statement is that Indian people had learned how to read the weather and a few were quite good at guessing when the rains would come. But Ray said also that no one is born with the spirit to call the rains anymore.

The burns are gone. And as the Northwest winters warm, the deep snows that keep the high-elevation berry fields free of trees will fall less frequently. Depending on our actions, they may disappear altogether. Without the snows, the berries in Indian Heaven will give way to forests of alder and fir. And just as steadily, as the Northwest summers warm, the glaciers that hang in the sky that are the source of the streams that nurture the salmon are melting away. April that year was the hottest ever on record for the Northwest, bringing scorching July temperatures deep into what should have been a cool, wet spring.

The rain fell all the way back to Portland and for the next four days, setting yet another new monthly climate record for the year—this one for precipitation in August. At home, I lay in bed at night and imagined the brown stream by our camp swelling, overflowing its banks with glacial melt from warm rain. In a different world, that ice would remain locked up by the high, cold air of the Cascade Mountains, a gift from nature to feed the wild salmon of our grandchildren.

Suppose Noah Had Replied Differently

God
Heaven   6,000 B.C.

Dear God,
    We are writing to let you know that we received your projection of an upcoming catastrophic climate change,

with accompanying flood. We also looked over your rather fanciful recommendation that we build and stock an "ark." As you know, climate forecasters remain uncertain as to whether it will rain for forty days and forty nights. Indeed, our own wise man has read the sheep entrails and projects a range that includes the possibility of no additional precipitation or perhaps even an extended drought.

Should heavy rains develop—and we assume, realistically, a likely worst-case scenario of seven days and seven nights—rest assured, the Noahs are prepared. We can quickly build two goat-skin rafts, on which we can take the entire family, our goats, sheep, chickens, camels, and enough seed to see us through the next season. Frankly, we are baffled by your "command" that "two-by-two" we take on board and feed all the animals of creation, including: serpents, mice, bats, horses, lice, buffalo, doves, elephants, geckos, frogs, chimpanzees, polar bears, moose, prairie dogs, quail, springbok, squirrels, mosquitoes, turtles, beavers, grasshoppers, and countless other useless creatures.

Really, what is the point? Most of these animals live thousands of miles away from human civilization. What use could they possibly be to anyone? And besides, do you have *any* idea what this ark would cost? In spite of your reputation for omniscience, we are quite frankly disturbed that you have been taken in by environmental extremists bent on wrecking our economy.

Sincerely,

Noah
Patriarch and Senior Research Fellow
The Noah-It-All Institute

P.S. What is a gecko?

CHAPTER 4

SPIRIT

True story: A young man once walked to the top of a snow-covered hill and paused, leaning his hand against the trunk of an old beech tree to catch his breath. He looked skyward, into the cold air. And there he saw that the winter sky, the whole fabric of the atmosphere, was held together by a fine but clearly visible gray mesh. He stood for a long time watching, for the first time in his life, the very weave of the universe. After a while, he looked down at the dried leaves poking though the snow on the ground, shook his head, and in his rational mind he understood that he was under the influence of some drug, mushrooms he had eaten earlier that day. But when he looked up again, still, the vision would not fade, and he stood there, entranced by the mysterious fabric that was the world above him, for a long time.

The young man's waking dream that day was not a spiritual experience; his vision was artificially induced. It does, nevertheless, illustrate the powerful *capability* of the human mind to draw profound meaning from the creation around us. This sense of heightened connection with the world, a sense of meaning, of purpose and guidance, of powerful comprehension, also emerges without drugs, naturally, in human spiritual experience.

In my own life, in periods of personal crisis, I have been drawn into the woods. There I found flashes of creative, transcendent insight. Among the trees, I have felt anxiety and despair that doubled me over until I prayed for help, and, as if by a miracle, had my questions answered. As I have grown older, I have come to know many people—mainline and evangelical Christians, Buddhists, Jews—for whom these altered states of spirituality are very real.

The human animal clearly has evolved with a spiritual capacity, a capacity to experience transcendent cognitive and emotional states and to gain insight and wisdom from prayer and meditation. This spiritual capability and impulse need not reflect a relationship to God; it can also be understood as an extremely cool feature of our evolution into *Homo sapiens*. Many people seek this kind of spiritual enlightenment through religious practice, mediated by organized communities, ritual, symbols, and sacred spaces such as mosques and churches, often familiar from childhood. Some of us seek it in nature.

There is a place that I go on the edge of a mountain, a place with flat sandstone outcrops that must have looked very much the same for tens of thousands of years. On hot, sunny days, the rock remains mysteriously cool. I lie there on my back, pressing my shoulder blades, the back of my head, the back of my hands, my spine, my calves and heels into the rock. I open my eyes and look skyward, and out of the corner of my eyes can see the lush and vibrant green of summer, the tops of ancient trees of oak and hickory. And in the blue above me, I watch the buzzards soar. I have done this dozens of times in my life, and sometimes, hypno-tized, I feel the Earth rise into my back. I feel the Earth breathe.

For many humans, the natural world is our sanctuary: church, synagogue, mosque. It is where we go to access those unusual, transformative mental states, to pray, and to seek and find a sense of connection, meaning, and renewal. Beyond the wealth and knowledge that the diversity of life brings, this is perhaps its most important meaning: It inspires our love.

Evolutionary Psychology, Spirituality, and Love of the World

People are creatures who can have waking visions, and through them experience a sometimes life-altering, sometimes fleeting sense of insight, joy, and deep connectedness with the natural world. Why do we, human animals, have this strange capacity?

In 1984, Harvard biologist E. O. Wilson proposed an evolutionary answer, in his pathbreaking book *Biophilia*. The title term means, literally, "love of life," and Wilson argued that a few million years of natural selection have instilled in us this powerful ability. Why? In a nutshell, people who grew up fascinated by nature, acutely sensitive to subtle shifts in biota, terrain, and weather, would be more likely to survive. In this way, Wilson argued, we all have evolved with an innate desire to become natural biologists. Each of us has the hard-wiring to allow us to passionately love interacting with life, because for most of human existence, those who were committed to learning the Earth's secrets, who obtained deep satisfaction from doing so, were more likely to reproduce and pass on that hard-wired capacity to their offspring.

Wilson offers an intriguing list of observations to support his hypothesis: a human preference for the savannah-like environments (think suburban lawns) that dominated human evolution in Africa; extensive evidence of the healing power of exposure to nature; and studies of the cognitive development of children's relation to the out of doors. Wilson's critics have countered that biophobia—fear of snakes, spiders, jungles, and wilderness, for example—is just as prevalent as biophilia (think of the quintessential New Yorkers, Woody Allen and *Sex in the City*'s Carrie Bradshaw). A more fundamental critique of the biophilia hypothesis is that what humans really are hardwired to love is not nature, per se, but complex surroundings. Lots of Americans prefer to spend time visiting shopping malls and video arcades rather than parks and streams.

A response to the biophobia critique is that biophobia itself confirms that humans maintain an instinctually deep—if complicated—relationship to nature. In many human cultures, snakes are in fact revered as powerful, and often associated with healing, while they are, simultaneously, feared. If humans have such a strong, complicated reaction to snakes, that is perhaps the best reason to keep them around: Such an unusual and powerful feeling is worthy of serious exploration. Snakes keep us on edge, and thus, alive.

In response to the complexity argument, Wilson would maintain that, even if we evolved to find joy in complicated surroundings, for two million years that environment was a natural one. Shopping malls are a poor substitute for forests:

> We are drawn to the natural world, aware that it contains structure and complexity and length of history as well, at orders of magnitude greater than anything yet conceived in human imagination . . . We need nature, and particularly wilderness strongholds. It is the alien world that gave rise to our species, and the home to which we can safely return. It offers choices *our spirit was designed to enjoy.* [emphasis added]

My point here is not to argue that biophilia is more widespread than biophobia, or that spiritual experiences gained in the wilderness are superior to those found in church. Rather, Wilson has provided a compelling argument for why, when many people begin to visit natural places, they do experience a sense of connection, and begin to care deeply for that environment, and often, for creation more broadly.

Wilson's work laid the foundation for the rapid growth of a field known as "evolutionary psychology." This approach seeks to explain human psychology broadly as a function of evolution by natural selection, primarily in the hunter-gatherer cultures that characterized the vast majority of human existence. The idea is that the human mind evolved to solve a variety of problems linked to survival: procurement of food, mate selection, child-rearing.

Attempts to understand mystical and transcendent episodes as either fitness-enhancing psychological adaptations or byproducts of evolution are not particularly satisfactory. Researchers have confirmed that spiritual experiences have a physiological basis. In one well-known experiment, a quarter of patients suffering from a particular type of epilepsy generating localized seizures in the left temporal lobe of their brains reported episodes of intense religious feeling. Recent work appears to have identified a particular gene that correlates with heightened spiritual sensibilities. And the facts that drug-induced visions resemble genuine spiritual experiences and that genuine spiritual experiences often flow from times of deep personal crisis reflect both common wisdom and established research.

Psychologist Leo Kirkpatrick, surveying this evidence, can only suggest that spiritual experiences are a poorly understood byproduct of other equally poorly understood emotional adaptations:

> Mystical experience probably involves the activation of brain region associated with some [simple, mechanical] function, that is ordinarily recruited by some more complex [psychologically adaptive] Functions. Hints about what these might be are suggested by other kinds of more mundane emotional experiences to which religious experiences have often been compared, such as awe, wonder, aesthetic experiences, creativity and other forms of "peak" experience.

Kirkpatrick argues that evolutionary psychology eventually will explain human mystical and transcendent states as byproducts of some more directly adaptive psychological mechanism. Others suggest that spirituality itself may serve an independent "Function": In some as-yet unspecified way, spirituality may promote individual or kin survival and reproduction. But again, the resolution to this debate is not my point. Whether one is an atheist, spiritual but not religious, or a believer in God, feelings of awe, wonder, powerful aesthetic experience, moments of transcendent insight, feelings of cosmic connection and peace, and even for some, mystical visions, are amazing and transformative pieces of the human experience. For many people (and in my view, as a result of biophilia) these spiritual elements of human life are facilitated through a deep love for creation.

God and Language

Given that, for whatever reason, humans are born with a capability to profoundly love and become spiritually connected to the natural world as they find it, the next step is to evaluate whether such love provides a strong moral basis for a defense of creation. There are two fundamental routes to such a moral stance: theistic and nontheistic.

For those who believe in God, the answer is a relatively easy yes (or no). Theists see layered on top of the evolutionary process relating humans to nature a divine presence, and from that, rely on their religious tradition to interpret right and wrong behavior. Economist Herman Daly along with ethicist John Cobb—both Christians—believe that

> each animal is immediate to God. Its suffering is immediate to itself and immediately shared by God. Those who love God will avoid causing unnecessary suffering even to the least of God's creatures . . . But whatever else God is, God is also the inclusive whole. The diversity of the interconnected parts of the biosphere gives richness to the whole that is the divine life. The extinction of species and simplification of ecosystems impoverishes God even when it does not threaten the capacity of the biosphere to sustain ongoing human life.

This belief, if strongly held, is clearly a sufficient basis for a vigorous political commitment to the fight for the clean-energy future that we need in order to protect the diversity of life on the planet.

The problem of course, is that not all Christians (or Muslims or Hindus or Jews) interpret God's will in this way.

Nevertheless, religious people who love nature have the powerful advantage of an immediate and familiar moral language in which to make their case. The writer Bill McKibben argues that the Book of Job

> will be to the emerging environmental theology what Exodus was to the theology of liberation. God's speeches from the whirlwind represent the first nature writing and probably the best. Since it was written, Job has troubled the rabbis and theologians because it is unlike anything else in the Bible. To me, it seems like a time capsule message, hidden in our tradition for this moment in time, designed to show us precisely the outlines of our current folly.

For McKibben, that outline involves man wresting from God power over nature. God says to Job:

> *Who gathers up the the storm clouds*
> *slits them and pours them out*
> *turning dust to mud*
> *and soaking the cracked clay?*

McKibben replies that, as a function of global heating, "increasingly, we do. The most recent studies show that extreme precipitation events—rainfalls greater than 2 inches in twenty-four hours—have increased 20 percent across this hemisphere." And if we do not think that will represent a "severe challenge to our understanding of what it means to be children of God," McKibben invites us "to go through the hymnal, crossing out the songs and stanzas that witness to God's power through thunder, wind, sparrow, whale, sunlight, springtime."

There is a growing religious-based climate movement, including the National Religious Partnership for the Environment, the

Eco-Justice Working Group of the National Council of Churches, the Evangelical Environmental Network, and the Committee on Environment and Jewish Life. These religious people have the moral authority, and ready language, of the Bible, Koran, or Torah to motivate and support their political efforts to protect nature. By contrast, those in the "spiritual-but-not-religious" camp lack access to this shared language. Moreover, Daly and Cobb raise a serious objection to Wilson's nontheistic biophilia as a moral basis for action: "One must admire the hard-headed logical consistency of the scientific materialist at his best. Wilson has followed logic where it has led, and that is to a dilemma: affirm transcendental value as a reality to which we can turn for guidance, or affirm the nihilism implicit in scientific materialism and give up all claim to truth or righteousness."

How can a person whose love of nature is a mere accident of evolution plausibly defend biodiversity against those who do not share that love? Mustn't agnostics fall prey to a paralyzing moral relativism? How can they summon the passion to fight for nature, if they recognize that they are doing so purely because they have been tricked into it by their genes?

Daly and Cobb highlight the critical importance of claiming "truth and righteousness," of being motivated by a strong sense of morality. If driving Northwest salmon to extinction is perceived by the human advocates for salmon as simply "sad" instead of "wrong," then salmon will disappear. Summoning the political will to slow our century's rush to extinction will require strong politics, and strong politics is driven only by powerful moral conviction.

How then can the atheist, or the spiritual-but-not-religious person, find the moral conviction that politics demands? The answer is not a hard one—really, they find it the same way that Christians like Daly and Cobb would: They look into their heart and, in so doing, "affirm transcendental value as a reality to which we can turn for guidance." The point is that the love, the sense of connection, that a religious person feels for nature is the same love that a spiritual-but-not-religious person feels. Nature is a mediating guide to spiritual experience for all of us, whether we think a particular manifestation of God is involved or not. The love is as deep and as intense, and if we truly listen to that love, then it leads us directly into a passionate defense of nature. If global heating threatens to wipe out half of life on this planet, and it is within your power to stop it, you stop it—even if you understand your love for creation to be "simply" the product of four billion years of evolution.

The real problem that nontheistic environmentalists face is not a depth of passion, but a failure of moral language with which to cultivate and nurture that passion. Here is an example chosen after searching four minutes on the web: an appeal to protect part of Alaska's North Slope as it appears on the web site of one environmental group.

From the sweeping vistas of the Arctic Refuge to the lush wetlands of the western Arctic, America's Arctic is a place of grand landscapes and rich biological diversity. But these places are threatened by oil and gas exploration and development that already sprawls across more than 1,000 square miles of once-pristine North Slope tundra . . . It is a deceptive land, harsh and unforgiving, yet surprisingly fragile. While the landscape at first appears barren and empty, the immense distances and the diminutive vegetation soon come into perspective and display a startling number and diversity of wildlife. White-fronted geese appear as if by magic among the tussocks; arctic poppies nod and dance in the wind; a long-tailed jaeger hovers just at the crest of a low hill before folding its wings and diving toward the tundra. In the distance, what looks like a scattering of stones turns out to be grazing caribou.

Another couple of paragraphs of this boilerplate descriptive prose leads to the clincher:

> A balance can be maintained between oil development and wilderness in America's Arctic, but only if the nation acts to permanently protect its most important wildlife and wilderness resources—the magnificent natural areas of the Arctic Refuge coastal plain.

This is an appeal for political action (or really, donations) devoid of any moral content. Why should I care about this ecosystem? I get some code words, generating a sort of low-level aesthetic concern: "rare," "pristine," "unique." But, really, so what? No one ever won a serious political battle advancing under the slogan "maintain balance."

This deadening language is partially the fault of environmental educators (like me), who emphasize the terminology of science and policy over that of ethics and philosophy. It is also the fault of environmental advocacy groups, who increasingly rely on "practical," utilitarian arguments to advance their policy agendas and who are nervous about deviating too far from the center of public opinion. Over the last few years, for example, a coalition of several major environmental groups have been puzzling out how to "message" to the public on global heating. Relying on extensive polling, focus groups, and communications consultants, the effort has turned up insights useful to activists (don't talk about the "greenhouse effect," talk instead about the "carbon blanket"). But the overarching message from the professional enviros is "don't scare people by telling them how serious the problem really is. They will just tune you out." Instead, focus on patriotic appeals to American innovation and technology, responsible management, and appeals to stewardship.

This kind of public opinion research is valuable when planning a national advertising campaign that must compete for the attention of consumers against commercials for cars and beer. And clearly, it is important to know where the audience you are addressing is starting from. Indeed, out of precisely this concern, even John Muir largely kept explicit references to God—though not spirituality—out of his journalism. But focus groups and polling cannot provide us with the language of leadership. Effective political communication comes from the heart, and the heart of concern about the impacts of humanity's climate destabilization is a spiritual connection to nature. Global heating, by unraveling the very fabric of the natural world, threatens to demolish that tie as no other project of humanity ever has. Quite simply, short of the devastation from all-out nuclear war, global heating dwarfs all previous threats to the integrity of creation that people have ever faced. If we cannot make that message resonate with voters, then we have no chance of stopping global heating.

While the de-spiritualizing of the language of environmentalism can be blamed partially on the professionalization of the field, it is more broadly a reflection of the moral relativism and secular cynicism of the broader culture. Many nonreligious people have grown uncomfortable with moral language. It is no accident that the most powerful spokesperson for the protection of the West's ancient forests in recent years, Julia Butterfly Hill, is the child of an evangelical preacher, as John Muir was before her.

> MUIR: The wilderness is full of plans and tricks to get us into God's light.
>
> BUTTERFLY HILL: When I entered the majestic cathedral of a redwood forest, I saw God as I had never believed possible—in the trees, in the ferns, moss, and mushrooms, in the air and water, birds and bears. I finally saw God with all my senses, with all of who I am, from the inside out.

Readers of this book likely will find Butterfly Hill's language on the one hand powerful and resonant. Yet I suspect some—

especially "environmental professionals"—might also find it vaguely embarrassing. The source of that embarrassment is indeed our moral relativism. If Julia is morally right, then the companies who clearcut old-growth forests must be engaged in actions that are morally wrong. But that is a hard judgment for some Americans to make. Who is to say that preservation is right and clearcutting old-growth is wrong?

The answer, of course, is that in a democracy, each of us has both the opportunity and the responsibility to say whether clearcutting ancient forests is right or wrong. This is not a simple choice, even for those who believe in the spiritual value of ancient forests: In this case, the livelihoods of our neighbors, timber workers, may be affected. Yet as citizens, we vote to define and defend our personal and community morality, by choosing political leaders who share our values, each day banning some actions and encouraging others. Only strong moral conviction supports strong politics. And unless passion about life on Earth is nurtured, and mass extinction is understood clearly in terms of good and evil, then political opposition to the great extinction wave of our generation will be weak and it will sweep across the next century unabated.

This is not to say that people fighting to limit global heating should become oppressive moralists. In general, politically committed people do more than enough finger-wagging. Rather, we need to become comfortable re-spiritualizing the language, acknowledging feelings of love, awe, and reverence, when we reflect on and discuss human relationships with nature. Here is Terry Tempest Williams describing the world just south of the same Alaskan coastal plain discussed above:

> My fear of flying in a small plane (I am riding on the back of a moth through clouds) is overcome by awe as we maneuver through majesty. These are not mountains but ramparts of

raw creation. The retreat of gods. Crags, cirques and glaciers
sing hymns to ice. Talus slopes in grays and taupes become
the marbled papers, creased and folded inside prayer books.

Overcome, awe, majesty, gods, sing hymns, marbled papers, prayer
books. These are words that speak to us all, regardless of the par-
ticulars of our faith. If forests are cathedrals and wild mountains
the ramparts of raw creation, then we will fight for them and we
will win. If they are instead a pristine, unique home to "wildlife
and wilderness resources," we will regret their passing and they
will disappear forever.

Talking about Right and Wrong

At the global-heating trainings that I help run, our participants go
through an exercise that seems quite simple, but is, in fact, remarkably difficult. They each develop a thirty-second "elevator
speech" that is a response to the question: "Why do you care
about global heating?" So your assignment now is to write five
sentences that capture your own personal, heartfelt response to
the question asked by your friend/co-worker/child/mother: "Why
do you care about global heating?" Try this in the space provided
before reading the rest of this chapter.

The typical elevator speech contains a list of potential scary consequences or else is a quick tribute to the job-creating power of clean energy. But the passion that would compel a questioner to engage further often is missing. The key to a captivating elevator speech is to spend some time uncovering why you care—*why you really care*—about global heating. How is climate stabilization different than or the same as, say, stopping AIDS or funding schools? Go deep, look into your heart, and discover what brought you to the point where you are spending your scarce spare hours reading about building a movement to support a clean-energy future. There inside, you probably will find a basic moral judgment: You know, or you strongly suspect, that human destabilization of the climate is just plain MORALLY WRONG. John Passacatando, Greenpeace Executive Director, once explained why he does what he does: Global heating, he said, is "an offense against grace."

Visit any web site or pick up any book on global heating, and you will read a lot about sustainability, biodiversity, intergenerational equity, pest and disease migration, crop loss, food shortages, and floods in low-lying countries like Bangladesh or Egypt. But the listing of threats, sounding vaguely like the twelve plagues, combined with appeals to bloodless abstractions such as sustainability and intergenerational equity, have no political traction; neither is this a language that can "sustain" (in an immediate sense) the passion of a movement. Our fellow Americans like to talk, not about the apocalypse, but about values, about what is right and wrong. If you try to convince your neighbors to be concerned about global heating based on some abstract principle, like "loss of biodiversity" or "impacts on agriculture," they will not care. Worse, they will recognize the lack of passion in your voice—and even perhaps—a lack of sincerity.

Whenever I give a talk about global heating, during the question-and-answer period, someone always wants to pin the blame on the greedy consumption habits of American consumers, their (not our) slothful ways, and their (not our) love of SUVs. My response is to ask them if they ever fly in an airplane? Because I do, a lot. Every mile traveled in an airplane emits, roughly, the global-heating-pollution equivalent of driving the same mile in a car. Flying from Oregon to Tennessee to see my dad or to distant conferences for work, I rack up tens of thousands of miles of emissions each year and am responsible for a lot more global-heating emissions than the typical American.

The point here is that while we absolutely need to talk about values, our focus needs to be on right and wrong policies, not right and wrong people. Government—our collective voice—consistently has made the morally wrong choices, subsidizing big fossil-fuel producers and failing to support clean-energy technologies. Not as individual consumers, but *as a political society*, we have failed to demand increased fuel efficiency from our vehicles, and as a result, we see our oil dollars fund terrorism in the Middle East, our young people suffer from an asthma crisis in our cities, and our planet's climate grow increasingly unstable.

If, instead of focusing on the need for change in Washington and Sacramento, Nashville and Columbus, or Austin and Lansing, we talk about our neighbors' lifestyle choices, most of us rightly can be charged with failing—miserably—to practice what we preach. More importantly, blaming instead of engaging our neighbors will get us nowhere. Put simply, global heating will not be solved by lifestyle changes or any shift to simple living, desirable as those changes might be. Today, and for the coming years, we need to pledge our lives, our creative energies, and our broad-ranging collective talents to creating a political movement powerful enough, in a few short years, to launch the new clean-energy markets that will mean the beginning of the end of the fossil-fuel era.

Why do I care about global heating?

"Where I live, in Oregon, global heating is going to cut the summer water supply in our streams and rivers by half, condemning my children to year after year after year of summer drought. Unchecked, global heating will kill more people and drive more animals to extinction than has any other industrial pollutant in human history. I think that's really wrong. Don't you?"

That's my elevator speech.

In sharp contrast, my friend Rhys Roth spends little of his time talking directly about global heating. Although Rhys works for an organization called Climate Solutions based in Olympia, Washington, he devotes his energy to building a partnership bridging the interests of renewable energy folks with farmers throughout the inland Northwest. Over the past five years, Rhys has patiently helped create an alliance that includes, among other organizations, the Idaho Hay Association, the Bonneville Environmental Foundation, Cascade Grain Products, the Confederated Tribes of the Coos, Lower Umpqua, and Siuslaw, Montana Ethanol Producers and Consumers, and the Oregon Environmental Council. The Harvesting Clean Energy Network now hosts an annual conference—most recently in Great Falls, Montana—that draws hundreds of participants. The focus at the conferences is on the nuts and bolts of making clean energy work for rural development—not on the threat of climate destabilization.

Sarah Severn talks more frequently about global heating in her role as Director of Corporate Sustainable Development at Nike. Working with the World Wildlife Federation, the company has set a serious global-heating-pollution reduction target for its owned facilities and travel; Nike is also analyzing the carbon footprint of its contract manufacturing facilities and has helped identify ocean shipping as a major target of opportunity for reducing global-heating pollution. As an advocate for environmental interests within a large multinational corporation, Sarah is an effective preacher of the "triple bottom line," pocketbook gospel that business can do well by doing good.

Jenny Holmes talks about global heating all the time. Jenny works for the Interfaith Network for Earth Concerns, and in a recent public hearing, had this to say to ultra-conservative Oregon legislators pushing a resolution to block state action on global heating: "This bill is not good for Oregon, it is not good for God's creation, it is not good for God's children, especially the poor . . . This bill is morally flawed. It promotes a view of the world that the earth's most powerful nation is not responsible for its actions. Is that the legacy that you want to be known for?"

Rhys and Sarah and Jenny are all devoting their working lives to enlarging the tent, building the political coalition that we will need to stabilize the climate. Obviously, we need to speak to our audience. But Jenny's argument, at the end of the day, is the one that carries us all forward. And it is the only one powerful enough to energize the kind of political change needed to preserve creation.

Morality, Fanaticism, Hypocrisy, and Politics

I have a two-year-old dog named Cricket—a shepherd-ridgeback-mutt mix with a gorgeous golden coat and a long, loping stride. Cricket is absolutely, one hundred percent biophilic: When I walk six miles on a trail, she will run sixteen. The woods keep her constantly engaged—smells, insect-laden tree bark, sights, fallen logs, creeks, high grass, new routes, sounds, swamps, the occasional squirrel or chipmunk. In nature, she is always poised and alert. Muscles and mind and senses working, she radiates joy.

Cricket and I are both mammals; we share something like three-quarters of our DNA. I have some of Cricket's joy in me, but in the everyday world of work and the city, it fades, and the vibrant edge of life fades too. Like Cricket, I go to the woods to hone that edge, for renewal. And that is the simple point of this chapter: Behind the utilitarian rhetoric of the movement to stabilize the climate are millions of people who share a deep and abiding love for creation, who believe in the redemptive, enlightening, enlivening, and spiritual power of the natural world.

This fact needs to be recognized, celebrated, and—put to use. Embracing the "truth and righteousness" inherent in protecting life from mass extinction is critical to invigorate the community fighting for a clean-energy politics. It is not merely sad that creatures are going extinct, it is wrong. It is wrong in this way: Creation as it has been handed down to us is a powerful foundation, a foundation of incomparable richness and complexity, for the realization of human spiritual potential. And it is wrong to deny these possibilities to the hundreds of human generations who will follow us. Many people believe this, yet it is very seldom said out loud. And by not speaking this truth, its vast power also fades.

However, at both the personal and political levels, taking a strong moral stand against mass extinction is not easy. Americans are uncomfortable with excessive public expressions of morality for two good reasons: We don't like fanatics and we don't like hypocrites. Fear of being labeled one of these often prevents people from expressing moral concern over extinction (more on this below). But there is another and bigger personal challenge: the pain of loss and of failure. If we embrace protecting species with a moral passion, then we open ourselves up to the love that we feel for them. When we read each day in the papers that another bloodline of our kin have winked out, the sadness, the sense of responsibility, can become immobilizing. How do we deal with these challenges?

First, fanatics and hypocrites. If I truly believe that mass extinction is WRONG, then must I, like Julia Butterfly Hill, spend two years of my life sitting in a redwood tree? Or worse, join PETA and devote my life to throwing red paint on runway models? Don't I have to eat 100 percent local and organic, stop using wood products in my house, give up driving my car (because oil production destroys sensitive habitat in the Amazon), definitely stop flying in airplanes, and so on, and so on? Doesn't acknowledging the morality of protecting nature commit me firmly to Earth First!ism: "NO COMPROMISE in Defense of Mother Earth!"? In sum, if I don't become a fanatic, then aren't I, by definition, a hypocrite?

Moral commitment and fanaticism are not equivalent. At the same time that I believe that protecting our habitat and stopping global heating are truly moral causes, I also hold a whole range of other, deeply felt moral beliefs: that all people have a right to health care; that we need to invest more in educating our children; that we need to fight discrimination. I also have two daughters to ferry around the suburbs, a job to do, family members to visit with, and I like to go to the movies and take the occasional urban vacation. The point here is that while people are indeed political animals, that is not all we are.

To be sure, stating that mass extinction is wrong and fighting it on those terms opens one up to the charge—by both friends and foes—of either fanaticism or hypocrisy. An important way to reduce the fear that these labels instill is to re-spiritualize the language of our concern for the future. This would render the moral perspective the more prevalent one, providing a stronger foundation for a committed and powerful politics. Along that journey, in the meanwhile, each of us knows inside whether we are approaching this calling in a spirit of humbleness and modesty.

A much more serious challenge is despair. In fact, as I write this, today's morning paper reports that "Farmers have over-run vast areas in the Congo's oldest national park, the latest threat to more than half the world's 700 remaining mountain gorillas." And I confess to a certain numbness that came over me on seeing the headline, as did a strong desire to quickly turn the page and skip the article. The depression I felt from reading this story, imagining the very few of Koko's people who remain alive in the wild, came back to me on and off throughout the week (and indeed, looking back on what I wrote, throughout the next year).

A young activist friend of mine, reading a draft of this book, liked the part that comes next about the fundamental role of politics, but said that he thought this spiritual chapter was a little silly. "Isn't it self-evident that what we are doing—fighting to protect life on the planet—is right? Why do we need to justify it in such 'irrational' terms?" I was buoyed up by his enthusiasm, but I also see many people even his age worn down by the grimness of the struggle. The fact is that the people who care deeply for creation seldom speak of that love, *even among themselves,* instead falling into a more socially acceptable, "rational" conversation about the benefits and costs of ecosystem and species preservation. But this is simply not a language that can nurture hope in the face of staggering loss. And staying open to hope and not going numb is critical to finding any meaning in the threatened life that surrounds us. This fight to slow the pace of mass extinction will be one in which we lose many high-profile battles and win only a few over the next few decades. But those few victories, if they reflect a change made by us in the political landscape, can save the world for hundreds of thousands of species.

Because this is a *political* struggle. Symbolic actions like Julia Butterfly Hill's two-year tree-sit have critical educational value; she wins minds and hearts. Lifestyle changes—buying a more fuel-efficient car, flying less, eating organic food—all these are important symbolic statements. They help each of us align our thinking with our actions and answer to external (and internal) charges of hypocrisy. But public education and modification of personal consumption habits have very limited potential to affect what will be the fundamental driver of mass extinction this century: global heating. And they certainly cannot achieve that goal in the critical next ten years. Only our national government has the ability to change the current rules of the game in ways that align underlying economic incentives with ecological sustainability—quickly enough, in the coming decade—stabilizing emissions of global-heating pollution and investing the billions of dollars we need in clean-energy solutions. Fighting to win this kind of power, real political power, is, finally, a way to find personal meaning in the beautiful diversity of life that remains in the world, and in the heartbreak of the daily disappearances that surround us.

CHAPTER 5

POLITICS

I grew up under the mythic shadow of the civil rights movement. In 1954, my New York City–raised parents left graduate school at Cornell, packed their old green Ford, and headed south, to the Cumberland Plateau and the town of Sewanee, Tennessee, population fifteen hundred. As they climbed the two-lane road to the top of the mountain, my parents came to a viewpoint, or really, two adjacent viewpoints, each providing a sweeping vista across the deep, shaggy coves spreading down to the flat farmlands lying in the valley below. A sign at the first viewpoint said "White View." The sign at the other said "Colored View." My dad had been offered a job teaching at a college that though small, bore a grandiose name and tradition: *The* University of the South. The job market was tight; my parents expected to stay a couple of years and move on. Fifty years later, my father, and my mother's ashes, are still there.

Ten miles down the road from Sewanee was a place called the Highlander Folk School. Founded in 1932 by Miles Horton, a southern labor organizer who had studied with Reinhold Niehbur at the Union Theological Seminary in New York, Highlander's original mission was to educate "rural and industrial leaders for a new social order." The school quickly became a major organizing center for the labor and later the civil rights movements, providing training, information-sharing, and networking opportunities. During the late 1950s, virtually every major civil rights activist, from Rosa Parks to Dr. King, attended Highlander workshops.

Civil rights was in the air as I grew up: My parents, with three other white families and four black families, brought suit to integrate our local public schools. My mother, along with her friend Mrs. Johnnie Fowler, the local NAACP chapter president, sat down for coffee at the town restaurant and broke the color bar there. The State of Tennessee brought trumped-up charges against Highlander; my father testified at the county court house in their defense. Highlander was shut down, forced to relocate to eastern Tennessee. Pete Seeger slept in our family room while on tour. My parents listened to records by Joan Baez and Miriam Makeeba. We sang "We Shall Overcome." And the signs came down. The South changed.

Here is the lesson I took. In America, political change happens in this way: People adopt a moral cause (abolition, women's suffrage, labor rights, civil rights, anti-war, anti-nuclear), and build a movement to educate the public. They demonstrate in courageous ways the depth of their conviction. They build a tide of moral sentiment that eventually converts even the *existing* political establishment. And then the movement's demands are codified in national legislation.

In college and graduate school, I was active in the issues of the 1980s: the efforts to stop apartheid, to end the wars in Central America, to freeze nuclear weapons production. For a while, though, my political interests took a back seat to a new job and family. Then, in 1999, with my kids growing older and my career established, I decided to apply the political lessons that I had learned to the central challenge of our time: global heating. And while the threat remained (at that time) somewhat distant and abstract, I believed that deep concern for the well-being of our children and grandchildren could form the heart of a new, powerful grassroots movement demanding an end to the fossil-fuel era and the beginning of a clean-energy future.

Some wonderful colleagues and I founded a nonprofit organization called the Green House Network. Our core idea has been to multiply leadership supporting the clean-energy revolution that we need to stop global heating—to be the Highlander Center of a burgeoning grassroots citizen's movement. As Highlander did during the labor and civil rights movements, we bring together citizen activists and educators, provide information, networking, tools, and organizing models. These leaders return home to engage in action and education—giving talks, organizing conferences, holding media events, meeting with political and opinion leaders—all helping to stop global warming.

Over the last six years, in partnership with great regional organizations such as Clean Air—Cool Planet in New England, the Massachusetts and Chesapeake Climate Action networks, the Blue Water Network in California, Climate Solutions in the Pacific Northwest, the Grand Canyon Trust, and the Environmental Law and Policy Center in the Midwest, the Green House Network has held a series of a sixteen weekend training workshops, with now over five hundred graduates. Collectively, the members of this group—including artists, engineers, retirees, clergy, nurses, students, scientists, stockbrokers—have talked to tens of thousands of people and have organized dozens of climate conferences and actions across the country.

I am very proud of the work that hundreds of Green House Network members have done. But eight years on, we are now further away then we were in 1999 to the holy grail of American single-issue political movements: substantive national legislation. In the climate change case, we remain in truly desperate need of a national commitment to develop the clean-energy alternatives that can forestall runaway global heating. When we started our organization, the United States was at least nominally committed to the goals of the international Kyoto global heating treaty. After President Bush's victory in 2000, however, he officially pulled us out of the Kyoto treaty, and no progress has come out of Washington since.

President Bush was not alone in his opposition to Kyoto or to action on global heating. He is joined by a large and powerful minority of the members of the House and Senate who share his ideological worldview. Over the last twenty years, a dramatic change has occurred in the political perspective held by the leadership of the national Republican Party. GOP leaders in Washington have largely abandoned their party's long-standing support for strong protection of the natural world, instead adopting a belief that above all—above all—regulations on businesses need to rolled back.

Global heating dwarfs all other environmental challenges that humans have faced, and yet, twenty years after the issue achieved mainstream recognition, the U.S. government still refuses to act. On reflection, the story I had inferred about American political change—from morally grounded grassroots movement to sweeping legislative accomplishment—was incomplete. I had left out the politics.

The Triumph of "Government is the Problem" Politics

The night Ronald Reagan was elected President of the United States, I was a twenty-year-old college student, wandering from bar to bar along the cold, windy streets of Madison, Wisconsin. Not very politically sophisticated, I nevertheless sensed a profound sea-change. The defining event of that day was less Carter's loss of the White House to a grandfatherly Hollywood actor, and more the surprising departure from the Washington stage of a generation of senator-statesmen. George McGovern, Birch Bayh, Frank Church, John Culver—these men represented the liberal wing of a thoroughly moderate, American worldview, shaped by the Depression and World War II. Their New Deal, "government and business as partners" ideology was as close to socialism as America would come.

That was not very close. In America, unlike in Europe, the market system was never fundamentally at risk from a labor-backed socialist movement. Instead, World War II raised a generation of pragmatic, relatively nonideological leaders, ranging from moderate conservatives to liberals, with a general faith in the power of America—via government initiative—to tackle serious social problems. Business was sometimes inconvenienced by the New Deal ideology but, even as major legislation piled up, not seriously challenged. Republican President Richard Nixon (following in the tradition of the great environmental President, Republican Teddy Roosevelt), oversaw the first big wave of federal environmental regulation in the early 1970s. At the same time, throughout the 1960s and 1970s, right-wing conservatism—as represented by Southern senators like North Carolina's Jesse Helms—was isolated, tainted by association with a discredited racism.

However, beginning with Republican Barry Goldwater's dismal defeat in the 1964 Presidential election, a new economic ideology infused conservative ranks: libertarianism. Inspired by the writings of the philosopher Ayn Rand and the economist Milton Friedman, young conservatives felt they had a better idea. Activist government, their story went, rather than solving social problems generally made them worse: Welfare payments bred welfare dependence; minimum wages led to unemployment; public schools led to poorly educated children. But more fundamentally, libertarians argued that the *real* social problem was not poverty, discrimination, environmental degradation, lack of affordable health care or decent education at all, but was, instead, *infringement of personal liberty by big government.* Government was "the leviathan," which, if not restrained at every turn, threatened to reduce free men and women to serfdom.

Libertarianism contains an interesting political contradiction.

On the economic side, its extreme laissez-faire philosophy harkens back to nineteenth-century, Victorian ideals of government, and pushes Republicans far to the right. However, on the social side, libertarian philosophy tends to align itself with pro-choice, anti-war, and civil-libertarian perspectives, characteristic of the left-wing of the Democratic Party. Far from foundering on this contradiction, however, the right in America has thrived on it.

The economic side of the libertarian ideology was perfectly tailored to a new generation of business leaders whose defining experience was not depression and national mobilization for war, but instead, increasingly global business opportunities and a growing thicket of frustrating state and federal regulation. And so the corporate money poured in. Beginning with the establishment of two free-market think tanks—the Heritage Foundation in 1974, followed by the Cato Institute in 1977—right-wing business executives and corporations devoted over a billion dollars to building the intellectual infrastructure that would support and then later staff the Reagan Revolution. Cheered on by right-wing media outlets such as the editorial page of the *Wall Street Journal* and the *Washington Times* (founded in 1982 and foreshadowing talk radio and then FOX News), in the space of ten years, Heritage, Cato, and a dozen similar young, conservative organizations, dramatically shifted policy debates on economic issues rightward, in a libertarian direction of tax cuts for business and the wealthy, privatization, free-trade, even free-er international investment, and deregulation at home and abroad.

Ronald Reagan embodied the essence of libertarian theory in his popular campaign slogan: "Government is the Problem." Throughout the 1980s, the Republican Party rallied around the slogan, with Republican moderates becoming an increasingly endangered species. Traditionally, moderate Republicans had been a leading force for strong environmental protection measures. Indeed, a Republican from Rhode Island, Claudine Schneider, introduced the first global-heating legislation in the U.S. Congress—in 1988! But by 1994, when the Republicans took control of Congress, moderate Republicanism had largely been chased out of Washington, and there was no room for an agenda that favored stronger government standards for pollution or serious investment in clean-energy technologies.

Nationally, while these anti-government and anti-tax themes resonated to some degree with the general public, few voters are motivated by a generic fear of government as leviathan. Nor was the public particularly excited about the economic agenda that was ultimately delivered: free-trade agreements, a stagnant minimum wage, budget cutbacks for schools and health care, and union busting.

By 2004 the "government is the problem" rhetoric itself was also ringing hollow. In a fundamental betrayal of the libertarian impulse that drove the right-wing revolution, the radical right, once in power, doled out huge increases in corporate welfare and lacked the political courage to carry through on their agenda of scaling back government, instead cutting only taxes. By the end of George W. Bush's first administration, a handful of large corporations were feeding at the public trough at levels never before seen in the United States, and the nation saw a trillion-dollar surplus dissolve into trillions of dollars of new debt.

Regardless of its policy successes or failures, libertarian economic thinking clearly provided the intellectual fire for the resurgent conservative movement. Most Republicans in Congress today are self-described "free-market conservatives." Yet as an ideology, libertarianism was too rarified, too dogmatic, and too divorced from the average person's experience to motivate members far-removed from the business elite. Right-wing politics needed a more visceral and more populist core. Enter—stage right—the second leg of the conservative revolution.

Reagan was carried into office, ultimately, not because voters were clamoring for small government or lower taxes on the rich. Rather, there was an angry backlash of swing voters against the ongoing cultural revolution that began in the 1960s. Commentators on the right exploited this tension and managed to convince an important block of voters that the overriding problem in America was precisely the *libertarian social attitude* shared by educated professional people like me: the so-called "permissiveness" of the so-called "liberal elite." Listen to Rush Limbaugh or Bill O'Reilly for twenty minutes and you are guaranteed to hear a story about some liberal elitest coddling criminals or pushing pornography.

Organizing in evangelical churches, social conservatives were remarkably successful in the political arena. The election of George W. Bush in 2004 capped a multi-decade run. In 2003, 45 U.S. senators and 186 representatives earned 80- to 100-percent ratings from the top three advocacy groups of the Christian right. Journalist Glenn Scherer argues that the anti-environment record of the Christian conservatives in office reflects an apocalyptic, "end-of-time" view of their electoral base. A more generous theory is that social conservatism thrives on a long tradition of American anti-elitism, and embraces the rural side of our historical urban-rural divide: Both positions translate into a (sometimes well-placed) distrust of environmentalists. Regardless, this anti-environment voting record of social conservatives meshes surprisingly well with the libertarian economic agenda of the anti-government corporate elite.

Recently, important splits in the conservative religious movement have emerged over global heating: Some prominent evangelical leaders have concluded that climate destabilization is destructive of God's creation, and should thus be opposed. Christian conservatives— much more than free-marketeers—are likely converts to the clean-energy agenda, precisely because they don't view government intervention supporting a moral agenda as bad, per se. As one leader put it: "We're not adverse to government-mandated prohibitions on behavioral sin such as abortion. We try to restrict it. So why, if we're social tinkering to protect the sanctity of human life, ought we not be for a little tinkering to protect the environment?"

Following staggering losses in the 2006 elections, the Republican Party faces serious internal battles. One of these soon will come to a head over global heating. On one side, traditional moderates (with a base in the northeastern and western states), farmers and ranchers, and pro-creation evangelicals are facing off against, on the other side, small-government true-believers allied with folks on the conservative side of the culture wars. With the "government is the problem" agenda in intellectual tatters, it may be that a major realignment around global heating will emerge. Until that happens, though, even following the 2006 elections, the vast majority of Republicans in Washington today pledge a genuine ideological allegiance to a worldview that simply cannot imagine a role for smart, ambitious government policy.

As I walked the chilly evening streets in the fall of 1980, the New Deal was dying, and the nation found itself engaged in an era of entrenched, ideological warfare, with the new hybrid breed on the right wing—the "government is the problem" social conservative—slowly gaining ground. Older attitudes hung on across the next decade; for example, substantive environmental legislation in the form of Superfund passed under Ronald Reagan. And in 1990, George Bush Senior signed into law the Clean Air Act Amendments, the last major environmental initiative to come out of Washington—now seventeen years ago. President Clinton's failed health-care reform initiative was the final gasp of New Deal liberalism. By 1994, the federal government had largely ground to a halt; no major progressive legislation emerged out of the eight-year Clinton tenure.

In the 2000 elections, the "government is the problem" political movement, which began marshalling its forces in 1964, finally broke through and seized real power. Consider the people with real influence over domestic policy in Washington during the first six years of the century: George W. Bush, Tom Delay, Dennis Hastert, Bill Frist, James Inhofe, Larry Craig, Antonin Scalia. These are not your father's Republicans. Conservative movement theorist and influential Bush advisor Grover Norquist summed up the philosophy that drives these men in a graphic way: "I don't want to abolish government. I simply want to reduce it to the size where I can drag it into the bathroom and drown it in the bathtub."

And yet the triumph of the movement was amazingly short-lived. A political leadership that believed that government was the problem managed to demonstrate that quite effectively, with spectacular failures at home (Katrina) and abroad (Iraq). And the American public responded, demanding accountability in the 2006 elections. Still, after building its intellectual and financial base for four decades, the "government is the problem" ideology retains a very powerful grip on the Beltway political establishment. Anti-government politicians who came of age in the right-wing bubble that was Washington, D.C., over the last decade will fight any ambitious clean-energy and climate-stabilization agenda tooth and nail. And so the critical challenge now is to win back the country—the U.S. Congress and the presidency—to solid majorities favoring a new worldview. We need commitment to a bipartisan American pragmatism that faces up to our real problems, and we need it in four and not forty years.

In almost all areas in which the country faces serious challenges—on war, terrorism, health care, civil liberties, reproductive choice, education, social security, the economy, tax reform—progress has been and always will be incremental and bad policies are reversible. But massive extinction is not. Each passing year leads to a silent but deadly build up of global-heating gasses in the atmosphere, an accumulation that is locking more and more species onto an extinction path. Each year lost deepens global dependence on fossil fuels, making it that much harder for us to change paths.

Hard, but still possible.

A New Progressive Era

While conservatives were busy reshaping the Republican Party behind a central, anti-government message, the Democrats fought
a rear-guard action, squabbling internally between centrists and
populists. During this period, politically engaged, liberal Americans often bowed out of electoral politics. Instead, liberal activists
organized around single issues—abortion rights, gun control, universal health care, gay rights, affirmative action, environmental
protection—motivating their constituencies with issue-specific,
often explicitly bipartisan appeals to "write their senators" and
donate dollars to D.C.-based lobbying efforts and lawsuits. In the
1970s and even into the 1980s, such a strategy paid off in terms of
occasional legislative and judicial victories. But by the 1990s, New
Deal and Great Society politicians were largely gone from Washington, and the "government is the problem" movement had elected
a sufficient number of hard-core ideologues to block any progressive initiatives.

For global heating, though, government is not the problem: carbon dioxide pollution is. Significantly slowing the rate of mass extinction will require three things. First, massive U.S. investment in
clean-energy technologies is needed to free us from dependence on
fossil fuels and stabilize the climate. Second, the U.S. government
needs to join with the rest of the developed world and cap emissions of global-heating pollution. Third, to keep carbon in standing forests and out of the atmosphere, major funding and support
for preservation efforts in developing countries is required.

These are all policies that require very active, very smart U.S.
government initiatives—very soon. Yet, no amount of lobbying,
no tidal wave of moral pressure, will ever convince a powerful
minority of key players in the Washington political establishment
to seriously support any of these policies. They are simply too
locked into an anti-government ideology to see beyond their blink-

ers. These politicians, and especially their worldview, have to be replaced.

It is a central point of this book that those of us who love the natural world and seek to preserve the diversity of life will get nowhere lobbying the existing Washington power structure for climate stabilization policy powerful enough to matter. This strategy generated incremental gains in the ideological world of the 1970s, and even the 1980s, but that world has been altered fundamentally. At the national level, moderate Republicans have all but disappeared and so has bipartisan compromise on global heating. The 2006 election created a shift in the balance of power in Washington, but it was not decisive. Voters cast their ballots on the basis of a botched war, not in support of a clean-energy agenda. Anti-government politicians retain significant power, and will block or water down any serious effort to tackle the global-heating crisis. Our efforts now must be directed toward building a dominating political power behind global-heating solutions in this, the most powerful nation in the world.

A clean-energy agenda is a way to stitch together many themes that resonate with Americans, building an exciting political coalition including moms, patriots, unions, and rural and religious voters, and environmentalists. Rewiring the world with safe, clean energy would address the asthma crisis in our cities, deprive Middle East terrorists of their source of funding, create millions of new jobs, provide new sources of farm income, stop oil drilling in sensitive habitat, and begin to stabilize the global climate.

One version of this vision is fleshed out in more detail in "The Apollo Project," proposed by a coalition of environmental and labor groups. The proposal demands a federal commitment to clean-energy initiatives similar to President Kennedy's launching of the Apollo moon expedition. This new Apollo Project could mobilize blue-collar workers behind the goal of stabilizing the climate, as it would generate good manufacturing jobs. The proposed ten-year, $300 billion program would help build more efficient cars, appliances, industrial motors, and public transit. It would develop renewable sources of electricity, with wind power and biofuels providing new sources of income for farmers in rural America. Apollo would create an estimated three million new jobs in the process.

Apollo reflects a deeply pragmatic response to the climate crisis. Leaving behind the stale left/right debate of "governments versus markets," Apollo asks simply, how can we move forward? Global heating demands that we use all the tools in our kit to promote a real clean-energy revolution, crafting a sustainable and prosperous future. This is not the time for some new New Deal, and certainly not for more free-market wishful thinking. Instead, we desperately need both the dynamism of markets—with their radical transformational potential—and a government smart enough to harness that energy in the direction of a climate-stable future based on clean energy and expanding forests.

What should we call this new worldview? We can certainly take inspiration from our own history. One hundred years ago, the American progressive movement took the country by storm, creating, for the first time in U.S. history, a major regulatory role for an active government: breaking up monopolies, regulating workplace and consumer health and safety, establishing, under state management, our unparalleled system of national forests and national parks. Progressives also democratized the political system, advocating and eventually passing reforms including women's suffrage, direct election of senators, and the replacement of party caucuses with the primary system. And progressive's achieved all this through electoral politics. The movement elected state legislators, governors, and presidents—Roosevelt and Wilson—from both the Republican and Democratic parties.

The original progressive movement was born as a reaction to unbridled corporate power and the excesses of the Gilded Age. In order to push through an agenda that reined in the abuses of the monopolies, sweatshop owners, and slumlords, the progressives had to tear down an ideological consensus that government had no right to interfere with the prerogatives of business. Teddy Roosevelt, never mincing words, declared that America had to "abandon definitely the laissez-faire theory of political economy and fearlessly champion a system of increased governmental control." Today, a twenty-first-century equivalent of the progressive movement needs to face off against the business-donors-first, anti-government ideology of corporate libertarians.

The clean energy movement can look to history for inspiration in other arenas. While the progressives of a hundred years ago had faith in the ability of government to be a positive agent for change, they also were political realists, acutely aware of the possibility of government failure. If the American public today seems turned off by an out-of-touch, over-bureaucratized, and occasionally corrupt government, one hundred years ago, reformers faced massive and pervasive corruption in the boss-dominated landscapes of city and state politics. In this context, progressives were focused as much on demands for government transparency and accountability as they were on activist public policy. In spite of the everyday reality they faced, progressives nevertheless saw tremendous power in the ability of government to serve the public interest.

In many ways, our task is much simpler than those of our forebears: the laissez-faire ideology pushed by the Heritage Foundation and the Cato Institute is much less entrenched in the American psyche (and the federal courts) than it was one hundred years ago. A uniquely American tradition of active government stretching from the Progressive Era, through the New Deal and the Great Society presents a model on which to build a new politics in which government plays an important and active role, setting the rules of the market game to help ensure strong communities and a sustainable world. And many government programs, such as Medicare and Social Security, are much more efficient in terms of over-head cost than their equivalents in the private sector.

Clearly, government action is not always, or often, the whole solution to social problems. But it is often an important piece of the solution. Government-set standards, subsidies, regulations, and taxes are critical levers in the tool kit of modern societies, and, especially at this moment in history, we cannot afford to throw them out the window in the service of a rigid free-market ideology. The planet is heating up at a rate far out of nature's bounds. Government needs to cap emissions of global-heating pollutants *soon*, and rapidly push the clean-energy solutions we are going to need in the next couple of decades. If we are to preserve the diversity of life on this planet, and as part of that process, build a strong, sustainable economy, government must take a much stronger role in some key areas than it now does in shaping the direction that our market system is headed.

A final lesson to take from the progressives is that real, substantive political change happens by transforming one or both of the major parties. The progressive federal agenda was initiated by Roosevelt the Republican and consolidated by Wilson the Democrat. When the Republican Party was recaptured by conservatives in 1912, Roosevelt bolted to form the Progressive Party. The Progressive Party did well, coming in second, splitting the Republican vote, and ensuring the election for Wilson. But the success of the effort was largely personal. Indeed, the party itself was better known by its Rooseveltian nickname: the Bull Moose. When Roosevelt refused the party's nomination in 1916, instead campaigning for the Republican candidate, the Progressive Party collapsed.

A modern movement powerful enough to stop global heating also will need to gain converts in both parties. The clean-energy agenda in fact appeals to a broad political spectrum ranging from moderates, to liberals, to farm-state conservatives who stand to benefit from development of biofuels and windpower, to religious conservatives concerned about stewardship of the planet, to security conservatives worried about oil revenues being recycled to terrorists. Indeed, clean-energy rhetoric is so attractive that it has been adopted widely by the Bush Administration. Here, for example, is the president in his 2006 State of the Union address:

America is addicted to oil, which is often imported from un stable parts of the world. The best way to break this addiction is through technology . . . To change how we power our homes and offices, we will invest more in zero-emission coal-fired plants; revolutionary solar and wind technologies; and clean, safe nuclear energy . . . We must also change how we power our automobiles. We will increase our research in better batteries for hybrid and electric cars, and in pollution-free cars that run on hydrogen . . . We will also fund additional research in cutting-edge methods of producing ethanol, not just from corn but from wood chips, stalks, or switch grass. Our goal is to make this new kind of ethanol practical and competitive within six years . . . Breakthroughs on this and other new technologies will help us reach another great goal: to replace more than 75 percent of our oil imports from the Middle East by 2025 . . . By applying the talent and technology of America, this country can dramatically improve our environment, move beyond a petroleum-based economy and make our dependence on Middle Eastern oil a thing of the past.

Sounds great. But no one was really surprised or even particularly embarrassed when, soon after his speech, Bush visited the National Renewable Energy Lab, to find that (surprise!) the previous week the Bush energy department had just laid off thirty employees. President Bush's FY 2007 budget did not come close to matching his rhetoric: He proposed cuts for research into wind power, energy-efficient hydro and geothermal energy (close to $100 million total), while boosting funding somewhat for solar energy and biomass fuels by the same $100 million or so—about the cost of two F-15 fighter planes. Hardly the stuff of a clean-energy revolution.

In the Northeast and California, moderate Republican governors such as George Pataki and Arnold Schwarznegger have carried the party's traditional torch—using government authority to protect the environment and push climate-stabilization legislation. At the national level, a few Republican "mavericks" (not moderates) like John McCain and Lindsay Graham understand how serious global heating really is. Some Republican farm-state senators—Gordon Smith, Chuck Hagel—are attracted to the potential for rural development from clean energy (while continuing to oppose global-warming legislation). But elsewhere in the nation, given the depth of its ideological conversion to libertarian economics, the Republican Party is finding it very hard to genuinely champion a serious climate-stabilization agenda.

If Republicans need a new generation of pragmatic leaders, Democrats need to get serious about delivering on a climate-stabilization program. National Democratic figures talk a good game, but the party has in recent years too often lost its nerve. Al Gore wrote *Earth in the Balance* in 1990, but the Clinton-Gore team got little done on global heating. The Kennedy clan generally have been environmental leaders, but in 2005 they worked to sabotage a plan to put a major wind farm off the coast of Cape Cod, because it would spoil the view. In 2006, citizen Al Gore returned to his roots with a vengeance through the release of his film, *An Inconvenient Truth,* and clean-energy rhetoric is politically fashionable. Once in power, though, will the Democrats have the guts to face down the oil, coal, and auto industries, as well as the right-wing noise machine, and lead this country into a serious response to global heating—stabilizing emissions within a decade, and investing tens of billions in clean-energy solutions? Not unless they feel the full weight of the American public demanding that leadership.

Rapid political change results from a combination of on-the-ground organizing and the winds of fortune. Teddy Roosevelt initiated the original Progressive Era when he ascended to the presidency unexpectedly following the assassination of the conservative McKinley. The brutal attack of September 11, 2001, could have provided a catalyst for a clean-energy revolution: My guess is that a President Gore would have used the event to promote American energy independence, appealing to America's innovative spirit to build the foundation for a secure and sustainable economic future. Instead, President Bush told us to go shopping, and bogged the nation down in a deadly, expensive, distracting, and ultimately self-defeating military campaign, the course of which will consume much of America's political attention for at least the rest of the decade.

By this measure, Naderism, in the year 2000, led not just to an unpleasant four-year bump in the road, but to a truly tragic lost opportunity for the effort to stabilize the climate. In a few parts of the country, the Green Party can play a useful electoral role at the local level, where Greens have a realistic shot at winning elections. But in our winner-take-all system, third-party politics at the state and national level can never amount to more than protest politics—with a very dangerous edge. Ross Perot threw the 1992 election to Clinton; Nader threw the 2000 election to Bush. September 11 provided the political cover for Bush to consolidate the "government is the problem" revolution. And in 2007, political decisions in America, the world's biggest global-heating polluter, remain largely in the hands of men who are openly contemptuous of global-heating science and who will see serious global-heating policy instituted only over their dead political bodies.

Many liberals have a long history of suspicion if not hostility toward the party establishments, and have been attracted instead to the more morally pure arena of protest politics. Protest politics, as a moral and educational force, can sometimes play a valuable, long-run role. But time, right now, is very, very short. Liberals, moderates, greens, conservatives, all people who believe in the ideal of progress, must come together to promote and elect candidates. We need to support candidates who will lead the fight for a clean-energy future from the most progressive districts; but we also must work just as hard, if not harder, for moderates in swing districts. The fate of hundreds of thousands of species on this planet may be decided in the next decade. To slow the rush to extinction, we need to achieve real, substantive political power, and we need to get there fast.

Politics is about both solid organizing and good fortune. Sometime over the next decade, Mother Nature may give us a dramatic wake-up call: a few years of severe drought in the lower Midwest; another unprecedented string of hurricanes along the Gulf Coast. At that moment, we will need to have the people in place—people of open minds and good faith—willing to rewrite the rules of the market and usher in a set of federal climate policies that can preserve much of the life of the natural world that we love.

State Success Stories

My home state of Oregon provides a nice illustration of both the evolution of American politics over the last few decades and the potential for breaking the paralysis that has resulted from the dominance of "government is the problem" ideology. To the rest of the country, Oregon is known as a progressive place: We have a sophisticated, statewide system of "urban growth boundaries," which have been very effective at controlling sprawl. We have hundreds of thousands of acres of wilderness. Portland, with its light-rail system and vibrant downtown, is a Mecca for urban planners. We also have a reputation as being socially liberal, epitomized by our "Death with Dignity Act," which permits physician-assisted suicide for terminally ill patients.

Yet, the same political dynamic that played out in the rest of the country is reproduced, in microcosm, in Oregon. Our innovative land-use laws, along with our bottle bill, are the legacy of a *Republican* governor, Tom McCall, who, in 1971, famously and politely asked visitors not to move to the state. Needless to say, Tom McCall Republicans have disappeared from the Oregon landscape, replaced by a homegrown version of the "government is the problem" Cato Institute (ours is called the Cascade Policy Institute), a social conservative movement that mobilizes around anti-gay initiatives and a local Limbaugh-equivalent, gun-toting talk show host. A unique wedge issue for us is forest management; the timber wars of the 1990s created a bitter urban-rural divide.

The state, like the country, appears in newspaper maps as polarized into red and blue regions: blue mostly on the coastal side of the Cascades, solid red on the interior east side. Rural Democrats, once a force in Oregon politics, largely have disappeared. Very few progressive statewide legislative initiatives have been promoted in the 1990s—none since 1993, when progressives lost control of the state senate. Oregon faces the same ideological warfare that plagues the rest of the country, with economic libertarians bent on union-busting and deregulation, especially of land use, and culture war conservatives seeking to impose social controls on sexuality. Our progressive image trades on policies largely established in the 1960s and 1970s; in recent years, a rising "government is the problem" dominance threatened to undo many of these policies.

Enter Jonathan Poisner and his team at the Oregon League of Conservation Voters. The OLCV was founded in 1972, and for its first couple of decades followed a typical post-1960s liberal strategy, focusing largely on public education. OLCV published a voter guide and operated an awareness-raising and fund-raising door-to-door canvas. The organization also endorsed candidates and made small political contributions, but was not much of a player in state politics. As Oregon's electoral landscape turned increas-

CHAPTER 5 POLITICS

ingly hostile to environmental concerns, however, the organization
saw the need to reinvent itself. In the mid-1990s, Poisner, building
on the work of his predecessor, set about remaking the OLCV into
a serious political organization. OLCV dumped their canvas and
began instead to build a real membership organization. Recognizing that their power was not money but people, OLCV organized
volunteer-staffed chapters, county by county.

The effort paid off. The county chapters interviewed and endorsed candidates, published local scorecards, and most importantly, organized get-out-the-vote campaigns across the state. Between 1998 and 2004, OLCV more than doubled its annual
budget, with much of that money going to pay staff supporting the
volunteers running the county organizations. In the 1996 elections,
OLCV fielded one hundred volunteers statewide. That number
grew to 350 in 1998, 1,100 in 2002, and 1,500 in 2004. In this
process, OLCV has managed to unite the environmental community in Oregon—single-issue folks working on salmon or sprawl,
forest health or clean air, transit or environmental justice—around
an electoral agenda.

And substantively, inch by inch, race by race, OLCV has helped
rebuild a clean-energy majority in the Oregon congress, winning
the state senate in 2004 and the house in 2006. This was not just
the result of OLCV's organizing efforts. Thanks to the hard work
of many progressive groups in the state (including my favorite,
"The Bus Project," www.busproject.org), in 2004 Oregon had the
highest voter turnout in the nation, at over 80 percent. And increasingly, it is not just environmentalists in Oregon who are facing
up to the magnitude of the climate crisis. In the fall of 2006, Oregon's *Capital Press*—"The Voice of Northwest Agriculture"—ran
a twelve-page spread on global heating, illustrating the severity of
the coming impacts of climate destabilization on farmers in our
region.

Oregon has not yet entered a new progressive era. Ballot initia-

109

tives over land-use laws and gay marriage continue to polarize the electorate. Even in this environment, however, OLCV has shown how, through good old-fashioned organizing, it is possible to build the kind of electoral base that can demand action to stabilize the climate. Poisner thinks that there is a lot more energy to be tapped. "There are 150,000 Oregonians who belong to environmental organizations," he said, "and we have only 5,000 members. That is a lot of potential volunteers."

Progress in Oregon galvanizing the green vote has paralleled success in other states. In 1997, just twelve states had leagues of conservation voters, and only four had full-time staff. Now thirty-two states have leagues, twenty-eight of these with full-time employees. Environmentalists for many years have been suspicious of—or uninterested in—electoral politics, content to support education and lobbying groups. In recent years, less than 1 percent of all grants to environmental groups have been directed toward organizations involved in electoral politics.

But over the last few years, people who care about the natural world—Republicans, Democrats, Independents—are at last beginning to wake up to the new political reality. Since the early nineties, the American Right has coalesced around a virulent anti-government ideology, and at both the state and national level has been running and electing candidates who share that worldview. To support a clean-energy future, we need to fight back and elect new politicians from both parties who truly understand the job that needs to get done over the next decade. As OLCV's Poisner reminds his volunteers: "Some people do politics, others have politics done to them."

In Oregon, now that clean-energy candidates have a majority in the state house, we have the power to pass laws that will help drive the clean-energy revolution. The first of these is the adoption of California's proposed "Clean Car" standards for cars and trucks. Unique among the states, only California has the legal authority to set its own emission standards for automobiles. Under a law passed in 2003, the auto industry will be required to reduce global-heating emissions in California from cars and light trucks by 25 percent and from larger trucks and sport utility vehicles by 18 percent. The industry will have until 2009 to begin introducing cleaner technology, and will have until 2016 to meet the new exhaust standards.

While only California can pass tailpipe emission laws, other states can then adopt them. In the summer of 2005, Washington State's legislature adopted the standards—conditional on the unusual provision that Oregon would do so as well! Following the 2006 elections, Clean Car standards in Oregon (and thus Washington) are now moving ahead with confidence.

Beyond vehicles, both California and northeastern states have passed state and regional policies to promote renewable electric production: power plant cap-and-trade systems for global-heating pollution emissions or "renewable portfolio standards" requiring a certain percentage of renewables in the electricity generation mix. The impact of the combined efforts of these states (along with Kyoto-signatory Canada) to promote renewable electricity production and cleaner cars could be dramatic. This half of North America produces more economic output than the three biggest European economic powers—Germany, France, and the U.K.—combined.

If there was a silver lining to the "government is the problem" sweep of the presidency, house, and senate in 2004, it was that people who believe in a serious clean-energy agenda had no choice but to pay close attention to their knitting. In the 2006 elections, the country reacted to the failures of the Bush Administration, but there is as yet no solid, clean-energy majority in either the house or senate in Washington. Given this, well-funded fossil-fuel lobbyists can stall serious legislation. Moreover, President Bush remains a firm believer in the notion that any regulation put in place to cap global-warming pollution will wreck the economy, and that government promotion of conservation, energy efficiency, and renewable power represent undesirable "market interference" (although, curiously, subsidizing coal, oil, and nuclear do not). The president's team will work to weaken fatally any global-warming legislation that crosses his desk. Some watered-down compromise may emerge, but a serious national program for stabilizing the climate will have to wait until after 2008.

With the federal government likely sidelined on climate policy until then, states remain a viable arena for clean-energy organizing. Working on statewide electoral campaigns, people concerned for the future of life on our planet are building the political bridges, experience, and infrastructure needed to elect candidates who will implement decentralized but coordinated state-level climate policy. Over the next few years, state and regional campaigns to promote wind power, biodiesel, and other renewable fuels, to adopt California Clean Car mandates, and to cap power-plant emissions will provide places to grow the movement to stop global heating. At the state level, working within *the existing political establishment,* people can and are winning important victories that give us heart. But, more importantly, in this arena we are building the political machine needed to elect the rock-solid clean-energy majorities necessary to make real progress at the state level, and within a few years, the national level as well.

What states cannot do is impose more rational, nationwide emission caps on power plants, fund research and development at the level we need to spur a rapid clean-energy transition, or negotiate international climate treaties that protect forests and the species that inhabit them worldwide. To seriously slow the process of mass extinction driven by global heating, state-level victories must lead—quickly—to federal electoral victories in 2008 and a powerful, bullet-proof clean-energy coalition in the U.S. senate and house by 2010.

The Bottom Line for Creation

And so we come at last to this book's profound little secret:

THE MEANING OF LIFE IS . . .
MAKING PHONE CALLS WITH A DOZEN OTHER
VOLUNTEERS IN THE DINGY STRIP MALL OFFICE OF
YOUR LOCAL CLEAN ENERGY CANDIDATE

I'm kidding, right?

No. In this moment in history, that is exactly where, for me, the meaning of life starts. For those of us who have been fortunate enough to develop true love for all creation, if we are also Americans, then, today, electoral politics is our ultimate calling. The only power in the world that can stem the wave of the sixth great extinction over the next decade is the vast economic and moral power of America. And the only way to harness that power is to control the policies set by the states and by the federal government of the United States.

This will happen only with a lot of phone calling of strangers. And going door to door; and organizing neighborhood caucuses; and raising money; and asking for volunteer time; and asking for votes; and founding campus clubs that promote political engagement by students; and training young political leaders; and building vibrant, statewide political (not single-issue) organizations and parties, committed to a clean-energy agenda. Bottom line: If you are one of the lucky few who have an extra $3,000 in your pocket, don't spend it on upgrading to a hybrid car or installing solar cells on your roof—instead, go to your local League of Conservation Voters and ask them which swing-district candidate needs the money, give it to her, and then give her your time, at least thirty hours. If you don't have $3,000, then give her your time, as much as you can give—and then give some more. When she is elected, demand legislation requiring that the government purchase nothing but high-efficiency vehicles for its fleet and install solar cells on every roof of every public school.

There are alternatives to a serious, full-on commitment to electoral politics. One can work for an environmental organization or school or business that is trying, desperately, to do the right thing: to save a patch of rainforest, to reduce pesticide use on campus, or to design a closed-loop industrial process that mimics nature. One can engage with the disappearance of species and ecosystems as an artist or journalist or scientist documenting extinction, extracting and storing disappearing knowledge, reflecting extinction back as a moral mirror for a saddened humanity. One can lead a low-consumption lifestyle, demonstrating a personal commitment to global-heating emission reduction. And through meditation and prayer and reflection one can, in the face of that very deep sadness, work to stay open-hearted and to love the world around us, even as, by the dozens and hundreds and thousands, the creatures we love go under, forever.

This is all work that must and will go on. But as Americans, we cannot fool ourselves into thinking that any of these actions—unless as direct means to the immediate end of electoral change supporting a clean-energy future—can slow the process of mass extinction. An hour spent on awareness-raising campaigns to promote voluntary recycling or bike-commuting or organic gardening or campus sustainability may be important for creating a stronger community. But in the face of massive global consumption of fossil fuel, it will also be a very precious hour lost. Toward the goal of preserving the diversity of life on this planet, the *highest and best* use of our time that really matters over the next decade will be working to elect clean-energy candidates, especially in swing states and swing districts across the country.

If, for Americans, politics is the only real way to preserve biodiversity, politics is also, in and of itself, good humanizing work.

Politics demands engagement with the human community in all of its complexity, insisting that we preach to the unconverted. Politics requires from each of us leadership, demanding that we develop a conception of "the good," and that we bend our skills as people towards persuading others that our view of the public interest is correct. It asks that, when given responsibility, we govern wisely, and that once in power, we face down the temptations that power brings. Ultimately, politics demands that we work as disinterestedly as possible for the good of the community, and by doing so, become the virtuous political animals that humans, at their best, can be. These demands are made equally on the president of the United States, and on the secretary of a local Republican or Democratic Party.

Of course, many politicians do not govern wisely or successfully resist the corrupting influence of power. In the spotlight of the media, the personal failings of political leaders are grist for daily headlines and cable talk shows. Rather than holding a reputation for wisdom and disinterested virtue, politicians are viewed, at best, as driven egotists, and at worst as venal reptiles. Disgust and cynicism are the fashionable responses to the American political process, breeding an apathy that cuts across all social classes. But by retreating from the human and ethical complexities of politics, we lose something important, become diminished.

Politics—electoral politics, in all its muddied, compromised glory—is the chore that Americans must undertake if we are to do anything substantive to preserve the diversity of life on this planet. But there is good reason to believe, as well, that a spell at politics is much more than just a chore: Indeed, political service itself always has been understood as a fundamental and critical part of the meaning of life. Facing the massive, ongoing destruction of the life we love, what more meaningful way is there to spend a life than fighting strategically and with the intention of winning for that solution?

Politics and Coal Plants

To stabilize the climate, one immediate, critical goal must be achieved, a goal for which national political action will be too slow. Over the next few years, a wave of proposed new coal-fired plants may be built. By 2012, new plants in the United States, China and India might be emitting an extra 2.7 billion tons of carbon dioxide—more than five times the cuts arising from the Kyoto treaty. These plants, if built, will last fifty to seventy-five years. Their construction will make it even more difficult to achieve a stabilization of carbon dioxide concentrations at levels that provide a margin of safety against catastrophic climate change.

The new coal plants that have been proposed are mostly old-style combustion plants. Using this type of conventional technology, it would be hugely expensive to capture the carbon dioxide emissions from the smokestacks. An alternative technology, called Integrated Gasification Combined Cycle (IGCC), converts the coal to gas prior to combustion; using this approach, it is possible to capture the carbon dioxide for potential sequestration deep underground. Industry likes to call IGCC a "clean coal" technology. A better name is "carbon-capture ready," since while the plants clearly can capture the CO_2, we don't yet know if we can successfully and cost-effectively prevent the captured carbon from leaking back into the atmosphere.

Will underground sequestration work well, or at all? It is too early to say. But new coal plants will be part of the world's energy mix for the next few decades, and sequestration must be explored as an option for dealing with emissions from these facilities. Given this, we must ensure, first, that very few new "conventional" plants get built. More importantly, the United States has to take the lead quickly in demonstrating the success or failure of commercial application of carbon-capture and sequestration. If it succeeds, then China and India will be able to design their new coal plants—which are undoubtedly coming, soon—to be carbon-capture facilities.

Given the current political constellation in Washington, the only way to stop the construction of dirty coal plants is state-level action. This is where a grassroots climate movement—using all movement tactics, from legal proceedings to peaceful protests—needs to grow, and grow fast. The clean-energy movement is a big tent, uniting farmers, people of faith, coastal residents, clean-energy entrepreneurs, students, and many, many others. Big tents by definition need powerful poles at their center. These poles must be built out of the strongest moral fiber. The civil rights movement had a vision of freedom nurtured by the Black Church; social conservatives have as a driving force their strong "pro-life" beliefs.

There is a moral vision at the center pole of the emerging clean-energy movement, a core that, if reinforced, can push our political process into achieving something akin to a miracle: an energy transition rapid enough to stabilize the Earth's climate within my daughters' lifetimes. It is this vision—our vision—that is truly pro-life. There is a very old and very powerful story that tells of our human obligations:

> I will establish my covenant with thee: and thou shalt come into the ark, thou and thy sons, and thy wife, and thy sons' wives with thee. And of every living thing of all flesh, two of every sort shalt thou bring into the ark, to keep them alive with thee; they shall be male and female. Of the fowl after their kind, and of the cattle after their kind, of every creeping thing of the ground after its kind, two of every sort shall come unto thee, *to keep them alive.*

After Noah and the family chose to build the ark and the flood waters receded, then God said:

> The rainbow shall be in the cloud; and I will look upon it, that I may remember the covenant between God and every

living creature of all flesh that is upon the earth; the waters
shall no more become a flood to destroy all flesh.

The rainbow sign was the symbol not only of a covenant between
God and humankind, but of a promise that binds God with humans and with *every living creature of all flesh that is upon the
earth*. This was a promise made at the very beginning of our collective story, a deep and important promise that is ours to keep.

Whether God-given or bestowed upon us as a glorious accident
of evolution, the human race today possesses both the intelligence
and the grace to do what we need to do to keep the global flood
waters at bay, to protect our animals and our insects, our birds and
our flowers, our trees, our brothers and sisters and our children. It
is our family's task, our calling, to see them—as many of them as
we can—through the coming storm. Indeed, through our collective, political decisions, we can lessen the fury of that storm. As the
actions of Noah's family preserved creation for us, to feed and
clothe us, to provide us with medicine and spiritual engagement,
and to share the Earth with us, so too can we for our descendants.

Growing from the depth of this moral vision, grassroots action
to stop dirty coal plants can play two key roles in climate stabilization. First, by demonstrating that the weight of the future is at
stake, a grassroots movement gains converts drawn to support the
central pole of the tent. But second, it reinforces the sense of deep
moral purpose that movement members experience and share.
Grassroots action is the kind of politics that attracts the energy
and enthusiasm and moral clarity of young people. It is the strength
of this moral compass that, ultimately, must provide the backbone
for the nations' politicians when they are called on to make the hard
social choices. We thus need a grassroots movement, but we need
a kind of movement that both can win real victories and that
doesn't alienate the political allies we need in rural communities,
in swing suburban districts, and in the business world.

Rallies in Washington are not the solution. A million marchers in D.C. merit about a minute of media coverage, support a boring parade of vaguely related speeches, and given the current dominance of the radical right inside the Beltway, simply would be ignored by the people in power today. A focus on Washington is premature. Before the national march in 1963, civil rights activists had been engaged in actions for a decade in cities such as Montgomery, Alabama, and Greensboro, North Carolina. Anti-nuclear and more recently anti-abortion protestors also built their base outside of Washington, with a focus on local actions. The grassroots clean-energy movement is growing in places such as Pueblo, Colorado, northern Wisconsin, and Billings, Montana. The target: proposed new coal plants. The demand: *at a minimum,* all new coal plants must be "carbon-capture ready."

The new investor enthusiasm for coal plants exists only because of "government is the problem" politicians in Washington. Investors are betting that they now have enough political influence to forestall global-heating regulations that will raise the costs of coal plants dramatically, or that, in the future, they can pass those costs on to energy ratepayers or taxpayers. Their model is the nuclear investors in the 1970s, who ignored escalating construction and waste storage costs, imposing huge burdens on today's utility customers and citizens. Our movement can make it clear to the investment community—with a commitment to a Ghandian/King style of grassroots movement—that we won't let energy companies play that game again. If plants are required to be carbon-capture ready, then carbon-capture costs will have to be integrated into investor planning, now. This is a basic principle of fairness on which all Americans can agree.

"Carbon capture ready" is a demand that is both ambitious and winnable. It is ambitious because it would help reveal the true costs of coal-fired power, raising the direct cost of new coal plants by around 20 percent. More importantly, by forcing companies to

119

acknowledge that they soon will have to pass the costs of carbon sequestration onto either ratepayers or stockholders, then investors, communities, and government regulators will pay closer attention to the hidden, long-run costs of coal. Recognizing these higher costs would encourage investment in clean alternatives, such as wind, solar, and biofuels.

The carbon-capture goal—because it must be won quickly—cannot wait for a national political change. It may be won through grassroots action. We don't need to go to Washington to force this kind of change. We can operate in our own back yards. Imagine hundreds or thousands or tens of thousands of people converging on the couple of dozen proposed dirty coal plants around the country. Imagine all of these people advancing a simple, doable, sensible demand: that America's leaders carry through on promises of technological leadership and deliver the cleanest available technologies to our communities. With this message and at these places we can stake our moral ground, grow our political movement, and win our future.

Progress and Passion

This morning, my e-mail in-box contained this news item:

> Hundreds of thousands of Scottish seabirds have failed to breed this summer in a wildlife catastrophe which is being linked by scientists directly to global heating. The massive unprecedented collapse of nesting attempts by several seabird species in Orkney and Shetland is likely to prove the first major impact of climate change on Britain . . . a rise in sea temperature is believed to have led to the mysterious disappearance of a key part of the marine food chain—the sandeel, the small fish whose great teeming shoals have hitherto sustained larger fish, marine mammals and seabirds.

Once again, "hundreds of thousands." To make a real difference for the creatures of the Earth, we must launch, in the next decade, a massive and fundamental transformation of the global energy system, and, through a comparably ambitious international effort, begin to put in place polices that protect the vast forest carbon-sinks in developing countries.

As Ross Gelbspan has pointed out, addressing the climate crisis is a way to transform the broader political and social environment: "Ultimately, a worldwide crash program to rewire the world with clean energy would, I believe, yield far more than a fuel switch. It could very easily, lead . . . eventually to a whole new ethic of sustainability that would transform our institutions, practices and dynamics in ways we cannot even begin to imagine."

I helped start the Green House Network on the premise that there were many, many people like me in the United States who were deeply worried about global heating, motivated fundamentally by a sense of responsibility to our descendants, and to the creatures of the Earth. I have not been disappointed. It has been truly a privilege over the last seven years to work with hundreds of citizen volunteers who have the passion to fight for a far-reaching, progressive political transformation in America. They give talks at their local Rotary Club, get themselves on cable access TV, organize delegations to visit the editorial board of their regional newspaper, wander the city square as a camera crew attracting homeless Santa Clauses, work hard for clean-energy candidates, and even run for statewide office. Born from their love of creation, their concern for their kids, and their commitment to global justice, these people share an overwhelming sense of urgency and a sense of mission. Theirs is a power that can usher in a new era in American politics.

The Progressive Movement of one hundred years ago laid the foundation for a uniquely American idea: Periodically, as new and serious economic and social contradictions emerge, the government needs to reframe fundamentally the rules under which market systems operate. Thirty years later, Franklin Delano Roosevelt's New Deal reforms adopted significant parts of the socialist platform, sharing the wealth and setting the stage for a half-century of political stability and economic prosperity. In the 1960s and 1970s, government policy started to pry open the doors of economic opportunity for women and minorities, and to clean up the air and water in our cities and towns. Today, America faces a new threat to economic prosperity and political stability, this time arising not only from the concentration of wealth and the abuse of power, but also, and more fundamentally, from a disregard for the constraints of nature. Once again, America needs to call forth her progressive spirit.

Close to a century ago, the labor leader Mary "Mother" Jones, following the murder of union activists, told her supporters to "mourn for the dead, but fight like hell for the living." This is advice for our time. We are in our own fight for the living, the remaining creatures of the Earth. Whether we win or lose will be important for our descendants, perhaps vitally so. But for each of us here day to day in a world where species slip away, the struggle to protect them—one of the very good fights on the Earth—is itself the meaning of life. Should those of us who love the diversity of life fail to organize and thus fail to gain real political power, then—regardless of our art, our journalism, our business efforts, our lifestyle changes, our prayers, regardless of all this—our children will suffer. The rush to extinction will proceed, tragically, unabated.

CHAPTER 6

SOLUTIONS

Pikas are hamster-sized mammals that live under rocks in high, alpine terrains. I met my first pika when I was eighteen when a friend and I rode a Trailways bus from near my home in the hills of south-central Tennessee to Boise, Idaho, disembarked, and got a ride in another friend's pick-up out to the Sawtooth Wilderness area. We spent ten days in jagged, granite mountains with high cirque lakes, strange country I had never seen before. In the cool evenings, from the boulder fields around the lakes, we would hear the curious, haunting, single-note cry of the pikas, like the sound of a penny whistle.

And then, two years ago, in yet another newspaper article, I read:

> Scientists believe that the American pika, a mountain-dwelling relative of the rabbit, is heading for extinction, and will one of the first mammals to fall victim to climate change . . . The animal lives between the tree-line and mountain peaks. As the climate heats up, it is having to go to higher altitudes to find suitable habitats . . . A study reported in the *Journal of Mammology* found that in pika populations at 25 places, nearly 30 percent (at 7 sites) had gone extinct. The locations are so remote that there seemed to be no other factor than climate change.

On Father's Day, 2004, my oldest daughter Emma and I joined a party climbing the south face of Mount Saint Helens. We hiked in and camped above tree line, in alpine meadows laced with delicate red and yellow wildflowers. The next morning we got up early and walked four hours up the snowfields to the ridgeline marking the lip of the volcanic crater, the sun shining along the final pitch. A wide snow cornice blocked our view down into the crater and the volcanic devastation to the north, but along the blue horizon we could see Helen's still-intact, snow-covered sister mountains: Rainier, Adams, Hood, Jefferson. We ate our lunch, and then we put cardboard in the seat of our pants, sat down on our butts, and took a wild 3,000-vertical-foot sled ride, thirty-five minutes back to camp.

On this Father's Day, along with my pack, I carried a heavier weight. As global heating raises temperatures in the Pacific Northwest, much of the snow pack that for at least one hundred thousand years has draped the Cascades will disappear forever, sometime in the next century. Even in the high mountains, winter precipitation will fall as rain, not snow. With the snow gone, the secret alpine meadows and their flowers will not survive, and summer stream-flow will diminish, driving more salmon to extinction. As a result of human pollution, the life even of the mountains has become transient. But in spite of that heavy knowledge, as I climbed the last hill into camp in the late afternoon, I paused, and then heard the penny-whistle cry of a pika. And my heart opened wide with joy. One moment of perfection.

I want this for my two daughters. I want this for the children that they may have, and for the generations of people who will live in the shadow of these mountains, forever. If we hold global warming to the low end—to 3, maybe 4 degrees—then somewhere in the Cascade mountains, there will be still be expansive fields of snow, blinding white in the June sun. The summer streams will still run full through alpine meadows. And these small animals of the mountain will remain, calling to human beings for thousands of years to come.

The philosopher Mark Sagoff challenges us to think about mass extinction in this way: "the destruction of biodiversity is the crime for which future generations are the least likely to forgive us. The crime would be as great or even greater if a computer could design or store all the genetic data we might ever use or need from the destroyed species. The reasons to protect nature are moral, religious and cultural far more often than they are economic."

Barring one of the catastrophic outcomes discussed in chapter 2, Sagoff's last sentence is probably right. But I wonder if future generations will spend much time dwelling on the habitat and wild communities we might have preserved for them. I am indeed immensely grateful for the great conservationists of the past, the Muirs, Roosevelts, Pinchots, Leopolds, and Browers who had the foresight to set aside public land in national forests and parks and later wilderness. Yet our grand- and great-grandparents could have done much more, and I don't spend much time cursing them for that. Just as current generations of young southerners are learning to love dogwoodless forests and some western kids hike pika-less mountains, members of future generations will adapt, and some will learn to love the natural world they inhabit—whether it contains a million fewer or half a million fewer species.

Ultimately, we cannot rely on the approbation of future generations to motivate us to protect the diversity of life on the planet; in many ways, they will have little idea of what they missed. Instead, I think, we could anticipate their immense gratitude from the wealth, the knowledge, and the rich experience that still remain in our power to protect.

Stop Global Heating, Part 1: Freeze and Invest

Humans are now engaged in an unprecedented natural experiment, in which we are altering the fundamental mechanism that governs the climate—and thus every single natural habitat—of the entire planet. It is up to the readers of this book, largely Americans, to decide exactly how far that experiment is going to go. Only 4 percent of the world's population, Americans emit 23 percent of the world's global-heating gasses. More importantly, our country has the technological know-how, the markets, the research centers, and the wealth, to quickly deliver the cost-effective alternatives to fossil fuels that can stop global heating. So do we, Americans, let the CO_2 concentration in the atmosphere, the blanket of fossil-fuel pollution that is warming the planet, thicken from where it today— 382 parts per million—to 450 parts per million (ppm)? Or 650 ppm? 850 ppm? Or 1,050 ppm?

Optimistic observers believe that we can stabilize atmospheric concentrations of CO_2 at 450 ppm. To get there, figure 2 shows what we have to do to our emissions of carbon dioxide. Rich countries need to meet the Kyoto targets by reducing CO_2 emisions to 5 percent below 1990 levels by 2010, and then cut emissions by a whopping 90 percent by the end of the century. Developing countries could keep increasing emissions until about 2040, but then they too would need to make significant reductions, to roughly half of 2000 emission levels by 2100.

Looking at figure 3, a person might be inclined to throw up her hands and go home. It seems we have before us an impossible task. But reflect for a moment on the primary pollutant from the transportation sector of one hundred years ago.

Got it? Horse poop.

We definitely have managed to reduce emissions of horse poop by more than 90 percent in one hundred years. To stop global heating, we need a comparable transformation of the energy sector: away from coal-fired electricity and petroleum-powered cars, to clean alternatives.

These clean alternatives are not a secret. The first step will be to create electricity from renewable sources: wind, solar, geothermal,

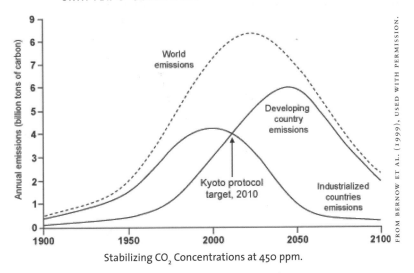

Stabilizing CO$_2$ Concentrations at 450 ppm.

FROM BERNOW ET AL. (1999), USED WITH PERMISSION.

biofuels. We can use that clean electricity to charge batteries to run our vehicles, or to make hydrogen to burn directly in cars, or run that hydrogen through fuel cells. We also can power vehicles directly from biofuels. There is an incredibly rich menu of technological possibilities. Rewiring the planet with these energy sources would provide millions of new jobs and lay the foundation for a sustainable and prosperous global economic system. The catch, of course, is that, with the exception of wind power and energy efficiency, these energy sources are currently too expensive relative to the fossil fuels that cause global heating.

The way to address this problem is also no secret: Government needs to invest in research and development and provide production incentives, so that as economies of scale kick in, these clean alternatives become cost competitive. My parents' generation showed the way. During the late 1970s, fearing oil dependence, the U.S. government provided subsidies to alternative-energy production. The state of California upped the ante with their own incentives. Investors rushed into solar and wind power technologies, and especially for wind, the policy was a tremendous success. Wind power prices fell rapidly: from $.25 per kilowatt hour in 1980, to $.05 in 1995. In 2007, with global capacity now close to 80,000 megawatts, in a good site and with access to transmission lines, wind power is the cheapest electricity source in the world, and it is the second-fastest-growing source of new power in the world, after solar. Wind power was a tremendous gift from my parent's generation to my generation: If they had not had the foresight thirty years ago to invest in these technologies, we would have no model for how to stabilize the climate.

My daughter Emma is now eighteen; within ten years, she may have a child and will be looking forward to sending him or her to college, and that will be 2036. I am forty-seven years old now and she will be forty-seven then. By that time, the planet will be heating up very rapidly, and Emma's generation will desperately need the tools to make cuts in global-heating pollutants on the order of 10 to 20 percent per decade. She will need low-cost fuel cells and low-cost solar arrays, cheap biofuels and geothermal technologies, and new tricks for using energy much more efficiently, if her generation is to stabilize the climate for her son or daughter. We know that these technologies take twenty years to mature; we know we need to invest serious money in them this decade.

Figure 3 also makes it clear why it so critical that the United States begin rapid development of clean-energy technologies today: to counteract the coming onslaught of global-heating pollution from China, India, and Brazil. Sometime in the next decade, developing countries will surpass rich countries, on an annual basis, as the main contributors to global heating. By 2040, clean-energy alternatives need to be cheap enough so that the Chinese will be installing wind farms and solar cells, not only because the planet will be heating up so fast, but also because electricity from these sources will be cheaper than from coal. The model here is cell phones: Countries like Ecuador are not stringing copper wires across their landscapes. They have gone straight to cellular technology. If America can deliver clean-energy options, developing countries can leapfrog our dirty development path, and raise living standards without compromising climate stability.

Where do we stand today with respect to meeting global-heating pollution goals? The European countries collectively appear on track to meet the Kyoto targets. The Japanese and the Canadians have signed onto Kyoto, and are making efforts, but will not meet the targets. The U.S. government is not participating in Kyoto, and our global-heating pollution levels are now 19 percent above 1990 levels.

George W. Bush pulled the United States out of the Kyoto Treaty for two reasons. First, he argued that it was unfair that poor countries were not required to cap emissions immediately. But once in the atmosphere, carbon dioxide lasts for at least a hundred years. So for the foreseeable future, the cumulative emissions of a century of CO_2 from rich countries will be to blame for the global-heating crisis. Beyond issues of moral responsibility, however, developing countries simply do not have the capability to develop and deploy the advanced technological solutions that will lead to a clean-energy revolution. In America, we do. And world leadership in the new generation of technologies that will power the world will be a solid foundation for a prosperous U.S. economy.

President Bush, however, also argued that we could not afford Kyoto, saying it would do "serious harm to the U.S. economy." Economists who studied Kyoto estimated, at the low end, that it might in fact cost America nothing, because we are so wasteful with energy. As an example, we could raise fuel-economy standards in our vehicles by over 10 miles per gallon and actually save money, because the higher upfront vehicle costs would be more than offset by the reduced fuel use (and we could create more U.S. jobs by sending fewer dollars abroad). Other economists argued

129

that there would be no free lunch, and put the costs of Kyoto at about $325 billion over the multi-decadal life of the treaty.

This is real money. However, between 2002 and 2006, the U.S. government spent more on the war in Iraq then Kyoto would have cost over its entire lifetime. By the beginning of 2007, the direct costs of the war were already over $350 billion. And because we put the expenses on the national credit card, every American family is on the hook for additional interest payments alone of well over $100 per year, forever, or at least until we pay off the debt. The Iraq war has been very costly to America in many ways, especially for the thousands who have given their life and the tens of thousands who have been severely injured. But from an economic point of view, the economy has not even shrugged. Was Kyoto affordable? Absolutely. Can we afford to stabilize in the next decade? Absolutely.

America has also learned some hard economic lessons by turning its back on Kyoto and the clean-energy future. In the early 1990s, driven by Jimmy Carter–era subsidies, California was the world's leader in both installed wind capacity and wind power jobs. But when U.S. policymakers lost interest, the industry shifted to Europe. In my home state of Oregon, when we go shopping for turbines, we don't buy them from California, we buy them from Denmark. More bad news came in 2006, when Ford Motor Company announced that it would lay off another thirty thousand workers, ceding the number-two automaker slot in the United States to Toyota. Starting twenty years ago, and understanding the way in which global heating ultimately would drive the marketplace, Japanese companies invested heavily in hybrid vehicles. Now they own that technology and the high-end jobs that come with it.

There is almost universal consensus among climate scientists that human-induced global heating is real. Less well known, many economists who study climate change also agree that the benefits of action to reduce global-heating pollution now outweigh the costs. While no one has taken any polls, a surprisingly diverse group of economists is calling for action. As early as 1997, several hundred economists—including eight Nobel prize winners—signed onto a public statement that read in part: "There are many potential policies to reduce greenhouse gas emissions for which the total benefits exceed the total costs."

More recently, the conservative American Enterprise Institute (AEI), in coalition with the centrist Brookings Institute, advocated a $10 per ton carbon tax on their web site. And then there is the *Economist* magazine. A recent issue features, against a backdrop of a glistening heap of coal, the banner headline "CO_2AL: Environmental Enemy Number 1." The accompanying editorial text reads: "The needlessly dirty, unhealthy and inefficient way in which we use energy is the biggest source of environmental fouling. That is why it makes sense to start a slow shift away from today's filthy use of fossil fuels towards a cleaner, low-carbon future." Not all economists supported Kyoto. Many would prefer a system of international taxes, rather than the hard-to-monitor cap-and-trade system that Kyoto endorses. And the $10 per ton carbon tax advocated by AEI and Brookings would not achieve Kyoto-level reductions. Nevertheless, these differences over policy details rest on an underlying consensus stretching across ideological boundaries. Weighing costs against benefits, most serious economic analysts agree that it is time to start reducing global-heating pollution.

The bottom line from an economic point of view is that the first steps to controlling global heating are really quite doable. If we are to hold warming to the low end, then the U.S. government must do two things this decade. First, we must put in place a national cap on global-heating pollution and freeze U.S. emissions. Second, we must begin to invest tens of billions of public dollars annually in clean-energy technology solutions. This has to happen now, because the technologies will take a couple of decades to mature and penetrate the global marketplace. If we don't invest now, it will be very costly for our kids to try to make the kind of pollution cuts that a rapidly warming planet will demand. It is clear that only the federal government has the power to mandate an emissions cap. But the federal government is also the only entity that can marshal the kind of research and development dollars needed to drive the clean-energy revolution as fast as it needs to be driven.

The market works wonders diffusing profitable technologies. But market actors face the short-term discipline of the stock market; very few companies can make large investments that have paybacks of longer than five years. In addition, the kind of fundamental R&D that is needed has broad public benefits, so that any company who did invest in revolutionary new technology and expected to be profitable in, say, fifteen to twenty years, would never capture all the returns. So the incentives are just not there for sufficient investment from the private sector. From jet planes to the internet, government support for new technologies has played a decisive early role in many major innovations. As the technologies mature and become profitable, then we can count on capitalism to do what it does best: spread the clean-energy revolution rapidly across the planet.

Freeze and invest. The very good news here is that both of these goals are readily achievable. California and the Northeastern states, not to mention Europe, Canada, and Japan, already are putting in place stabilization plans. The money needed to drive a clean-energy revolution amounts to a couple of months of spending to support the war in Iraq. The obstacles are neither economic or technical. Again, we come back to politics.

Stopping Global Heating, Part 2: Maintain Standing Forests

Ending dependence on fossil fuel is one leg of a climate-stabilization strategy needed to protect the diversity of life on the planet. The other is a global effort to protect forests that store massive quantities of carbon in their soils and biomass. Globally, forests have shrunk by about 50 percent since humans began clearing land for agriculture ten thousand years ago. Only fifty years ago, old-growth forests covered 40 percent of the ice-free land; that figure is now less than 27 percent and rapidly declining. According to the United Nations, in recent decades forests worldwide are being clearcut at the rate of about 1 percent per year. The forests that remain are increasingly fragmented, isolating species and breaking up the migration routes that both plants and animals will need to find new habitats as the planet heats up.

The first three bullet points in biologist E. O. Wilson's set of proposals to protect global biodiversity call for forest protection:

- Protect so-called "hot-spots" rich in species, many of which are in tropical rainforests.
- Keep intact the remaining "frontier forests"—again southern rainforests, as well as boreal forests of Alaska, Canada, and Siberia.
- Cease all logging of old-growth forests everywhere.

Yet forest protection in developing countries has proven very difficult. Given the current rules of the global market economy, poor people in developing countries simply cannot afford to keep their forests intact. Logging, farming, and ranching all put immediate food on the table. Ultimately, the only way to achieve forest-preservation goals is to make large-scale forest preservation pay. And the only way to do that is to create markets for the vast carbon-storage capacity found in these forests. Developing countries are providing people in rich countries real economic value if they preserve their forests. A comprehensive global climate-stabilization package is the only realistic way to ensure that poor people who depend on forests get paid for that service.

The good news is that the climate crisis provides a unique opportunity to make this happen. Global heating demands global solutions. Rich countries must cut emissions of global-heating gasses. But steady economic growth in the developing world means that, eventually, those countries will have to cut pollution too. Developing countries, however, will not reduce emissions if doing so compromises in any way their efforts to reduce the desperate poverty faced by so many of their people. Is there a way through here?

We saw above that the first step is to drive down the costs of clean-energy technologies, so that in a few decades, cutting emissions will be low-cost way to produce power. But it is also true that preserving the carbon storage that forests provide in developing countries will keep a devastating pulse of global-heating pollution from entering the atmosphere. Rich countries will need to pay for that forest-protection effort. A global climate-stabilization program would support clean-development projects in the developing world; for example, Brazil or Indonesia might attract funds to build markets for sustainably harvested wood or to promote farming and ranching practices that lead to no net loss of forest lands. Canada and Russia might also gain credit for protecting their vast boreal forests.

In this way, stabilizing the climate not only would protect biodiversity directly from the direct threat of rising temperatures and changing rainfall patterns. By placing a dollar value on the carbon-storage benefits of the world's forests, a climate pact also would provide the resources that countries need to protect critical habitat across the planet. Only the effort to stabilize the climate has the potential to mobilize rich countries to spend tens of billions of dollars to protect Southern forests, because only the climate crisis can mobilize those resources on the basis of immediate, Northern self-interest.

Ross Gelbspan, a journalist who has confronted the magnitude of the climate crisis more honestly than just about any other observer, argues for a global currency transactions tax raising $300 billion per year to fund the diffusion of clean technologies in poor countries ("wind farms in India, fuel cell factories in South Africa, solar assemblies in El Salvador, and vast, solar-powered hydrogen farms in the Middle East"). The fund also could support sustainable management of old-growth forests, providing credits to people in poor countries who support themselves from standing forests— thereby maintaining globally valuable stores of carbon.

Gelbspan concludes: "Given the magnitude and urgency of the accelerating pace of climate change, the only hope lies in a rapid and unprecedented mobilization of humanity around this issue." His strategy for "unprecedented mobilization"—or something like it—is what our governments must lead in the next decade. Stopping global heating by means of a clean-energy revolution and the global protection of forests obviously is a *huge* social project that will test the limits of human ingenuity and creativity. But one hundred years ago, visionary political leaders from the Progressive Era established a system of national forests and parks in our country that are the envy of the world and are today the treasure of an entire nation. Why not a similar, global vision for our generation? Also beginning a hundred years ago, the world embarked on a massive transformation of the energy system, from horse drawn carts to automobiles, a transformation that was largely completed in forty years. We need to cut that time in half.

New Politics

Given the power of "government is the problem" ideology in the United States today, serious climate policy may seem like a pipe dream. But it is critical to remember that it is only politics that keeps us from moving ahead. Cross the U.S. border into any other industrial country in the world, and the debate shifts from "should we cut emissions?" to "how fast can we cut emissions?" British Prime Minister Tony Blair: "On climate change, we need to build on Kyoto but we should recognize one stark fact: even if we could deliver on Kyoto it will at best mean a reduction of 1 percent of global heating [emissions]. But we know . . . we need 60 percent reduction worldwide. In truth, Kyoto is not radical enough."

Scientists tell us that climate change is very real, with very serious consequences for billions of people and natural ecosystems across the planet. Economists largely agree that we can afford to cap emissions now and begin investing in the technology solutions that today's college students will need in twenty years. Public opinion, by majorities of around 70 percent, backs government efforts to fight global heating. And business opinion—IBM, Johnson and Johnson, Boeing, Toyota, Nike, Shell Oil—all these companies have stated publicly that global heating demands a new set of rules for the economy. The month before I wrote this sentence, John Browne, CEO of British Petroleum, told the UN Council on Foreign Relations in New York: "It would be too great a risk to stand by and do nothing." Fighting climate change later, when it becomes a serious problem, instead of now, while there's still some chance that it could be controlled, Browne continued, could be "so disruptive as to cause serious damage to the world's economy."

It is as if the only people in the world over the last few years who don't get how serious the climate crisis is work inside the Beltway. Listening to our leaders in Washington, it seems we have learned nothing from decade after decade of smaller environmental crises—from burning rivers, killer smogs, thalidomide babies, Love Canals, and ozone holes. Only this time, we are taking our stupidity from the local to the global: we are playing an incredibly dangerous game of dice with the entire planet's climate control system, and our officials in power can only parrot the party line of the "government is the problem" think tanks: "No, we can't stop this, it will cost too much."

As I have given talks across the country about clean-energy solutions to global heating, however, it has become clear to me that unlike so many of our senators and representatives in Washington, most Americans get it. They get that this whole discussion of costs is mostly just a smoke screen for inaction. They get that it will be our children who will pay the price for that inaction. And they get, more fundamentally, that creatures who foul their own nests on the scale that humans are doing it today are asking for big, big problems. But these nascent insights have not—yet—translated into any serious demand for change. There is growing awareness in America, there is growing concern, but what we really need now is:

FOCUS

For the last year, I have been directing a major national educational initiative on global heating, called Focus the Nation (www .focusthenation.org). This project was motivated by the early 2006 statement of Dr. James Hansen, the top climate scientist working for the U.S. government, quoted in chapter 1 of this book. It is worth reading again:

> How far can it go? The last time the world was three degrees [C, or 6 degrees F] warmer than today—which is what we expect later this century—sea levels were 25 meters higher. So that is what we can look forward to if we don't act soon . . . How long have we got? We have to stabilize emissions of carbon dioxide *within a decade,* or temperatures will warm by more than one degree. That will be warmer than it has been for half a million years, and many things could become unstoppable. If we are to stop that, we cannot wait for new technologies like capturing emissions from burning coal. We have to act with what we have. This decade, that means focusing on energy efficiency and renewable sources of energy that do not burn carbon. *We don't have much time left.* [emphasis added]

Americans do not know this. We need to talk about it. At this moment in history, we owe our young people at least one day of focused discussion about the momentous decisions that we either will make or fail to make over the next ten years. As a nation, we need to discuss what needs to be done to secure a prosperous and sustainable future.

And so, on January 31, 2008, over one thousand colleges, universities, and K–12 schools across the country will hold simultaneous, one-day educational symposia on the topic of stabilizing the climate in the twenty-first century. At each school, a team of five or six organizers (faculty, students, and staff) will engage twenty to thirty of their colleagues—teachers, students, alumnae, community members—to participate as educators for Focus the Nation. We expect an average participation of two to three thousand students on each campus. All this means five to six thousand organizers and twenty to thirty thousand educators engaging two to three million young people in a serious, academic, nonpartisan, far-ranging discussion of the critical climate-stabilization decisions that either will be made or not, over the next decade.

Focus the Nation is set to occur at the very beginning of the 2008 political primary season, just after the New Hampshire and South Carolina primaries. Across the country, Focus the Nation events each will end in a political round table, with organizing teams inviting local, state, and national elected officials and other decision-makers to come to campus for an evening discussion (not a partisan debate) about actions to stabilize the climate. Imagine three thousand political leaders each receiving dozens of invitations to speak in their district or town—on the same day—about a clean-energy future.

Focus the Nation takes no stand on policy issues or particular legislation and each campus will decide the direction of its event. We offer only the following hypothesis for national discussion. If the United States seeks to help hold global heating to the low end of the forecasted range, then during the next decade the country must do two things: Freeze emissions of global-heating pollution and begin investing tens of billions of dollars annually in the development and commercialization of the clean-energy technology solutions that we will need to stabilize the climate.

Whether they accept or reject this hypothesis, participating schools each will set in motion during the fall of 2007 a deliberative process on their campus. Using as tools articles in school newspapers, speakers, debates, and on-line forums, each institution will begin a discussion of the policy options facing the nation in the coming decade. On Focus the Nation day, January 31, 2008, each institution will vote online on their top three or four priorities for national action. The project thus challenges educational institutions to engage their students in serious discussion, but more than that, as citizens of this nation, to draw their own conclusions and make them known to America's leaders.

Building from a base in educational institutions, Focus the Nation is incorporating participation by faith groups, businesses, and civic organizations. Churches, mosques, synagogues, and temples will be holding their own educational events; businesses will be doing the same with their employees; and civic organizations ranging from rotary clubs to bicycle clubs and garden clubs will also be participating. The goal is to spark a truly national discussion about clean-energy solutions to global heating, and more fundamentally, to focus the growing and widespread American concern about climate destabilization into a unified voice for national action.

At the college where I teach, Lewis and Clark, a group of twenty or so faculty, students, and staff spent some time during the summer of 2006 planning the outlines of our Focus the Nation event. We set three goals: do justice to the complexity of the issue (there is no magic solution); engage every academic department on campus (psychology, religious studies, theater, economics—everyone has something to say about the challenge we face); and involve students, alumnae, and the community as educators. Throughout the day we will hold plenary sessions in our largest venue, the gym, concurrent with classes. Although there is no mandate to participate—Focus the Nation is not a "teach-in"—we anticipate that the entire campus community of faculty, students, and staff will attend many of the sessions. Along with schools all over the country, in the evening we will invite local political officials to come and talk with us about a vision for the future.

At the end of our discussions at Lewis and Clark, we came to an important realization. Not only will our institution be doing this, but so will Portland State, the University of Portland, the Portland Community College campuses, Reed College, and Pacific University. High schools and middle schools throughout the region will be participating as well. At least a dozen churches, synagogues, and mosques will be holding their own events, with businesses and civic organizations involved too. Someone suggested, let's rent the convention center for the evening and bring everyone in the region together to hear some music, share ideas, and try to reach some collective consensus on our future. This is the most exciting thing about Focus the Nation: This same conversation gets reproduced wherever teams start planning their own events, from Idaho to Virginia to New York.

As I write this chapter, Focus the Nation is only a couple of months out from launch. We have been talking about the project at colleges and universities and conferences across the country. Everywhere, the reception has been enthusiastic. Over 450 institutions already have signed up to participate, with dozens more committed, and many college presidents have endorsed the effort.

If you are reading this book in the year 2007, then I hope you already have heard about Focus the Nation. I hope you are helping build a Focus project at your campus or in your community, creating a national dialogue so big that it cannot be ignored. If you are reading this book after January 31, 2008, I hope that Focus the Nation was in fact what we are now dreaming that it can be: a watershed moment when the American conversation about global warming shifted from fatalism to determination to face the challenge of our generation. But even if it does achieve (or has achieved) all that it can, Focus the Nation will be only one day.

What is more critical is what happens the day after. Unless the energy and excitement that comes from this national discussion is translated into political commitment by millions of American, we will not be able to take the serious steps that we need to stabilize the climate. Politicians, when faced with an organized citizenry, will always say that they will do the right thing. But unless Americans stay politically focused on a clean-energy future, the hard choices simply will not get made, and the world will not stop getting hotter.

And so beyond focus, this is a moment in history that demands a new politics. For some of us, this will be a politics grounded primarily in a concern for our children and grandchildren and for the people across the globe who will suffer from global heating. For others, it will be motivated as well by the clear recognition that the diversity of life on Earth is at stake. Those of us fortunate enough to have fallen in love with the natural world have only a short window of time to change the course of history. And that passion— for pikas and polar bears, for salmon and seals, for chimpanzees and cheetahs and flowers and corals and fish and frogs and trees and tortoises and bees and birds—must carry us into a hard, serious, and sustained fight to gain the real political power to preserve creation.

We need as well to pursue a politics that clearly recognizes the enemy. The obstacle to climate stabilization is not the apathy of voters or the wasteful habits of American consumers. It is instead a self-consciously manufactured, anti-government ideology that has paralyzed our nation. This ideology must be countered by a new pragmatic worldview. Quite simply, we need smart, aggressive action from government, to cap emissions of global-heating pollutants and push a new generation of clean-energy technologies to the edge of commercialization, so that they will be there when our children need them.

We need, finally, to focus on real politics, electoral politics. It is time to start going door to door and convincing our neighbors to vote for a clean-energy future. This is steady, sometimes unrewarding work. But in America today, we have the unique, historical privilege of doing that work. Unlike our progressive forebears, or those in many other countries of today, we are not threatened by dogs, firehoses, blacklisting, firing, beating, torture, goons, imprisonment, or death squads. We are free, if we choose, to make our own future.

CHAPTER 7

FIGHTING FOR LOVE

On a beach, in the distance, a woman steps out from the ocean up onto the rocky shore. She is in a wetsuit, a striking feminine form now standing silhouetted by the midday light reflecting off the ocean. In her right hand, she holds a heavy bag. On this island, women keep a tradition of diving to collect the fruits of the sea.

I am with the woman I love, and we walk down from the top of the rocky beach to a place where a tent is set up. The diver too has walked to this place. She has handed over her catch to the other women gathered there and now squats powerfully beside them. The diver's face, framed in her black neoprene hood, is deeply, deeply wrinkled. She has been diving from these rocks for sixty years. Knives are busy as the women peel the harvest from their shells, slicing the gathered stalks of life, filling a stainless steel bowl with shapes of opaque pink and brown and green, still shining with the salty coat of the sea.

Men are just up the rocks, eating. And my companion, urgent in her desire, wants some, so she buys bottles of rice wine to share around, and the men hand us metal chopsticks, and me a full shot glass. I sample; she savors, greedy, startles once, when a small creature heaves in its smooth, brilliant shell as she carries it to her lips.

Watching, some small thing moves in me, deep in my heart.

And one hundred feet away, the ocean is changing.

Burning coal and oil and gas to power more than a century of economic growth, humans have pumped vast quantities of carbon dioxide into the air. Global heating has been one result, as the carbon blanket in the atmosphere has become thicker and thicker. But recently, scientists have discovered that not only the atmosphere, but the oceans too have been absorbing vast quantities of that carbon dioxide. This CO_2 is reacting with the ocean waters and creating acid. You and I are now acidifying the entire ocean. The pH of the ocean surface is now 0.1 units lower than it was in the pre-industrial era. If my daughter visits this beach when she is that diver's age, the pH could be another 0.3 units lower. In the next two centuries, global-warming pollution could drive the oceans to an acid level not seen in 300 million years.

Acid waters kill coral reefs, which serve as nurseries to much of the ocean's diversity. Shallow tropical reefs are already directly threatened by global heating from bleaching in hotter seas; now these corals, along with deep and cool water reefs face this second threat of acidification. But there is an even larger fear.

Most of the animals we ate on that beach—mussels, sea cucumber, mongay, abalone—live lives protected by shells, spun out of the chemical compounds in the sea water. Many of the tiny invertebrates at the base of the marine food chain also have shells made from calcium carbonate. As the ocean becomes saturated with CO_2, the acid waters might begin to dissolve these shells, killing off the food sources at the bottom of the chain, and having a devastating impact on marine ecosystems all the way up. According to one prominent researcher, "many of the marine species we rely on to eat could well disappear. In cartoon terms, you could say people should prepare to change their tastes, and switch from cod and chips, to jellyfish and chips."

Stabilizing the climate is not about saving the planet. The planet will be fine. If half of the creatures whose beauty shines in our world disappear in the next century or two, give the Earth a few million years. The magic of evolution will repopulate the world with new and fantastic trees, flowers, mammals, reptiles, insects.

Stabilizing the climate is not about saving the human species. Even inhabiting a vastly diminished creation, and on a much hotter planet, in one form or another, humans will carry on.

Stabilizing the climate is a precious opportunity to pass on to all future human beings gifts of immense value, gifts that, once gone, will be beyond the imagination and skill of humanity to recreate.

We stand at a moment in history without precedent. Decisions that are ours to make over the next ten years will have a sweeping impact on the future direction of life on the planet. None of us asked for this. Twenty years ago, global heating was to me and most of us only a science fiction fable. And yet suddenly, our generation has been called upon to prove of what vision humanity is capable.

When I walk in the forest or fields with my mate, my love, she has a curious habit. Whenever she sees an unfamiliar fruit or flower or root or berry, she reaches out to caress the part of the plant that has attracted her, and asks me, asks the world: Can I eat this? She is an urban gal, not a botanist, but she knows that there are many wonderful things to eat in the woods and in the sea. And there is nothing, nothing more beautiful than watching this woman, nut-brown arm reaching out to touch a food that neither of us has known before.

In our century of extinction, no matter what each of us does with our lives, many, many creatures in this world will disappear. But, if we build a new politics, many, many other creatures will survive to inhabit and bless the world for generations to come.

NOTES

Chapter 1. The Century of Extinction (pp. 1–22)

Glacier retreat in Montana is projected in Hall and Fagre (2003). The *Boston Globe* story on seals is Nickerson (2001). Data on rate of species loss is from Wilson (2002: 99; 58–61), while Wilson (1992: Chapter 3) discusses the Cretaceous extinction. The 2004 Nature article is Thomas et al. (2004). The UN temperature projections are from IPCC (2001). The "global heating" recommendation is Lovelock (2006). UK government report on economic impacts is Stern (2006).

Sources for information on individual extinction threats are as follows: wildflowers: Connor (2003); frogs: USGS (2004); lions: British Broadcasting Company (2003); polar bears: Morrison (2004); turtles and tortoises: Kirby (2003); chimpanzees and gorillas: Hirsch (2003); United States plants: Deenen and Rembert (1999); corals: Buddemeier et al. (2004) and Young (2003).

Paleontological evidence on extinction can be found in MacPhee and Sues (1999). Data on forest cover is from Wilson (2002: 99; 58–61). O'Neil and Oppenheimer *Science* article is from 2003. Hansen quotation is from Hansen (2006).

For a good discussion of salmon in the Northwest, see Lichatowich (2001). Muir is quoted in Dennis Williams (2002: 195). Dogwood blight is discussed in Thins and Evans (1997). A representative vegetation model for the United States is Bachelet et al. (2001).

Chapter 2. Wealth (pp. 23–44)

The list of ecosystem services and quotation are from Daily (1997: 3–4, 10). The Keynes quotation is from 1972: 325, 331. Trends in global and regional economic output are from Maddison (1995). Falling UN population projec-

147

tions are discussed in Eban Goodstein (2004: Chapter 20). For an introduction to Herman Daly's economic perspective, see his textbook with Farley (2004). The peak oil story is told in many places, for example, Goodstein (2004: Chapter 19).

The best source for global warming science, including levels and projections for CO_2, is the IPCC 2001. US emissions information is from US Energy Information Agency (2006). Northwest impacts are described in Wolf (2005). Impacts on the Rimac River are from Scully (2005); global water shortages from Coonan (2006) and Stern (2006). Abrupt climate change scenarios are discussed, for example, in Hansen (2005).

The sustainable development definition is drawn from World Commission on Environment and Development (1987). The *Science* article on valuation is Costanza et al. (1997). The quotation from an economist is from Polasky et al. (2005a). The Hudson River story is told in Daily and Ellison (2002: Chapter 3). For more on Portland's urban growth boundary, see Goodstein and Phillips (2000).

Analyses of habitat preservation in Malaysia, Cameroon, Thailand, and Canada are discussed in Balmford et al. (2002). Eban Goodstein (2004: Chapter 8) discusses existence and option value, and economic valuation of non-market goods in general. Data on bioprospecting are from Polasky et al. (2005a).

Wilson is quoted from Wilson (1992: 347–48) and Wilson (2002: 106, 108). The statistics on the spread of high-yielding varieties, loss of natural diversity, seed banks, and the Irish potato famine are from Imperial College (2003). Rice and maize stories are from Heal (2000: 11–13). Information about Iraq is from Clarke (2003). The Council for Agricultural Science and Technology is quoted in Heal (2000: 13). On mosquitoes, see Gorman (2004). For an overview of the inconclusive "web of life" debate, see Kareiva and Levin (2003). The Willamette Valley study is Polasky et al. (2005b).

Chapter 3. Knowledge (pp. 45–66)

Autumn's work on geckos is described on his web site at http://lclark.edu/~autumn/dept/geckostory.html; quotations are drawn from the media sources listed there and from a personal communication. The extraction "red list" can be accessed at www.iucn.org. Roosevelt is quoted in Dalton (2002: 183). Statistics on human and animal waste and chlorine treatment are found in WHO (2004) and Environmental Defense (2004). Garfield's headquarters is described in Teriault (2002). The EPA analysis of Living Machines is US EPA (2002). Hawken's principles are discussed in (1995: 12).

Koyukon information is drawn from Nelson (1986: Chapters 8 and 10). Quotes are from Nelson (1986: 164, 175, 202–203, 238). Debate over human

origins and behavioral traits in relationship to bonobos and chimpanzees is discussed in Wrangham and Pilbeam (2001: 5–6 and Table 2); primate sexuality is discussed in Judson (2002: 152–65). Details on Koko are found in Patterson and Matevia (2001) and Patterson and Gordon (2001). A discussion of research on aging in apes is in Erwin et al. (2001). Statistics on the bushmeat trade are from Patterson and Matevia (2001). Polar bear research is reported in Morrison (2004), and Arctic study quote is from Spotts (2004).

Barber's thesis on cultural homogenization is found in (1992). Nelson is quoted from (1993: 20). Norton quotations are from (1987: 211–13). Cross-cultural study quotes and information are detailed in Kahn (1999: Chapter 9).

Chapter 4. Spirit (pp. 67–87)

This chapter presents a theory of justice based on "moral sentiment" following Locke (1980) and Hume (1966), in which reason and culture uncover and mediate an underlying, God-given or evolutionarily hard-wired sense of justice. For nonspiritual moral arguments in defense of the natural world and in opposition to global heating, see Ralston (1994) and Donald Brown (2002).

To head off some confusion concerning evolutionary psychology, here are two things that evolutionary psychologists do not believe: First, people's personalities and actions are not determined by their genes. Our genetic code provides a suite of mental hardwiring that determines our ultimate capabilities. But "nurture" is critically important for determining which hard-wired frameworks are put into play and even how they are structured as we grow. Moreover, environmental factors build the software programs that ultimately will run on the physiological hardwiring. Thus, culture is a critical determinant in human development.

Second, evolutionary psychologists do not claim that human behavior itself should be understood as promoting "survival of the fittest." People do lots of crazy, self-destructive things. Instead, humans have inherited a suite of psychological mechanisms that generally are adaptive for passing on their genes—love, lust, reason, loyalty, guilt, status-seeking—and these psychological adaptations play out within a given cultural context. For example, one human adaptation is love for a mate. Cross-cultural and biological evidence suggests that as a species, we are on the monogamous end of the spectrum, although the evolutionary reason for this is not well understood. Jealousy as an emotion may have arisen as a way to enforce monogamy. And yet jealousy can lead to highly maladaptive behavior, such as murdering one's spouse and offspring in a jealous rage.

The Wilson quotation is from (2002: 48). The Epilepsy experiment is described in Kirkpatrick (1999); quotes are from page 940. The potential genetic relationship to spirituality is discussed in Hamer (2004).

Daly and Cobb quotes are from (1989: 387, 392). Job discussion is from McKibben (1999: 1197–98). Sierra Club web site description of the Arctic coastal plane is found at http://www.sierraclub.org/wildlands/arctic/northern_slope.asp. Muir quote is from Dennis Williams (2002: 195); Butterfly Hill quote is from her book (2002: 2). Terry Tempest Williams is quoted in (2004: 28). Passacatando and Holmes quotes are drawn from personal communication.

On the DNA similarity between dogs and humans see Paulson (2004). The report on mountain gorillas is from the *Toronto Star* (2004).

Chapter 5. Politics (pp. 89–122)

Highlander's historical mission is taken from their web site: http://www.highlandercenter.org/a-history.asp.

The story of the rise of the radical right, from Goldwater on, has been told in many places; a fun start is Dionne (2004). Corporate largesse in the Bush administration is detailed at Slivinski (2004). Statistics on the voting records of senators and representatives, are from Scherer (2004). Evangelical leader (Rev. Rich Cizik) is quoted in Laurie Goodstein (2005). Norquist quote is from an interview with Mara Liasson on National Public Radio's *Morning Edition,* from May 25, 2001.

The Apollo Project is described at http://www.apolloalliance.org/. Roosevelt is quoted in Bailyn et al. (1985: 613).

The text of President Bush's State of the Union address can be accessed at http://www.c-span.org/executive/stateoftheunion.asp. NREL layoffs, and 2007 budget figures are discussed in CNN (2006) and Union of Concerned Scientists (2006). Klein (2006) reports on the Kennedy's and Cape Wind.

The OLCV story, and quotes from Poisner are from a personal interview. Oregon's *Capital Press* ran their global warming special issue on Friday, September 22nd 2006. Statistics on the composition of environmental grants is from Denise Ryan at the National League of Conservation Voters, personal communication, November 18, 2004. California's CO_2 tailpipe standards and other state-level initiatives are described in Hakim (2004).

Projected coal plant emissions are from Clayton (2004). The cost premium for IGCC is discussed in, among other places, Swope (2006).

The crash of seabird populations is described in McCarthy (2004). Gelbspan (2004: Chapter 8) outlines his solution; quotations are from pages 200 and 204.

Chapter 6. Solutions (pp. 123–142)

Pika information is from Paul Brown (2003). Sagoff is quoted in (2005). U.S. emissions are discussed in Vidal (2005).

The windpower story is told in Eban Goodstein (2004: Chapter 20). Kyoto compliance by region is reported in Suzuki Foundation (2006). The President's quote on Kyoto is from Bush (2001). For a low-cost Kyoto scenario, see Koomey et al. (1998). On fuel economy standards, see the National Academy of Sciences (2002). On the cost of Kyoto and the Iraq war, see Sunstein (2006). The 1997 Economists Statement on climate change was organized by Redefining Progress in San Francisco; it is on their web site at http://www .redefiningprogress.org/programs/sustainableeconomy/econstatement.html. The AEI position paper is American Enterprise Institute–Brookings Joint Center for Regulatory Studies (2001). The *Economist* citation is July 2004.

Data on forest cover loss is from Wilson (2002: 99) while his agenda is outlined in (2002: 160–64).

Gelbspan is quoted in Gelbspan (2004: 204–205).

Blair is quoted in Gelbspan (2004: 127). Browne is quoted in *Globe and Mail* (2004). Hansen quotation is in Hansen (2006).

For sample public opinion results, see Langer (2006). Businesses supporting action on climate change include the members of the Business Environment Leadership Council at the Pew Center on Climate Change. For more on Focus the Nation, please visit www.focusthenation.org.

Chapter 7. Fighting for Love (pp. 143–145)

Ocean acidification data are taken from Orr et al. (2005); and Calderz and Wickett (2003). Fish and chips quotation is from Dr. Carol Turley, the head of Britain's Plymouth Marine Laboratory, quoted in McCarthy (2005).

REFERENCES

American Enterprise Institute–Brookings Joint Center for Regulatory Studies. 2001. "Market Mechanisms to Slow Global Warming," Hundred Million Dollar List. Washington, D.C.: AEI-Brookings Joint Center for Regulatory Studies.

Bachelet, Dominique, Ronald Neilson, James Lenihan, and Raymond Drapek. 2001. "Climate Change Effects on Vegetation Distribution and Carbon Budget in the United States." *Ecosystems* 4: 164–85.

Bailyn, Bernard, Robert Dallek, David Davis, David Donald, John Thomas, and Gordon Wood. 1985. *The Great Republic.* Lexington, Mass.: Heath and Company.

Balmford, Andrew, Aaron Bruner, Philip Cooper. 2002. "Economic Reasons for Conserving Wild Nature." *Science* 297 (August 9): 950–54.

Barber, Benjamin. 1992. *Jihad Vs. McWorld. Atlantic Monthly* [3 March] 269–63; 53–65.

Bernow, Stephen, Karlynn Cory, William Dougherty, Max Duckworth, Sivan Kartha, and Michael Ruth. 1999. America's Global Warming Solutions. Washington, D.C.: World Wildlife Fund.

British Broadcasting Company. 2003. "Lions 'Close to Extinction.'" BBC News Service, September 18, http://news.bbc.co.uk/2/hi/science/nature/3119434.stm, accessed January 2, 2005.

Brown, Donald. 2002. *American Heat: Ethical Problems with the United States Response to Global Warming.* Lanham, Md.: Rowman & Littlefield.

Brown, Paul. 2003. "American Pika Doomed as 'First Mammal Victim of Climate Change.'" *The Guardian,* August 21, 2003.

Buddemeier, R. W., J. A. Kleypas, and R. B. Aronson. 2004. *Coral Reefs & Global Climate Change: Potential Contributions of Climate Change to*

153

Stresses on Coral Reef Ecosystems. Arlington, Va.: Pew Center on Global Climate Change.

Bush, George W. 2001. Text of a Letter from the President to Senators Hagel, Helms, Craig, and Roberts. White House News Release, March 13, 2001. http://www.whitehouse.gov/news/releases/2001/03/20010314 .html accessed March 00, 2001.

Caldeira, Ken, and Michael E. Wickett. 2003. "Anthropogenic Carbon and Ocean pH." *Nature* 425 (25 September).

Clarke, Tom. 2003. "Seed Bank Raises Hope of Iraqi Crop Comeback." *Nature* 424 (July 17).

Clayton, Mark. 2004. "New Coal Plants Bury 'Kyoto.'" *Christian Science Monitor,* December 23, 2004, 1.

CNN. 2006. "Bush Blames Lab's Funding Shortfall on Mix-up: Employees Rehired before President's Visit to Tout Energy Plans." Tuesday, February 21, 2006; Posted: 2:07 P.M. EST (19:07 GMT), http://www.cnn .com/2006/POLITICS/02/21/bush.energyfunding/, accessed August 17, 2006.

Coonan, Clifford. 2006. "Global Warming: Tibet's Lofty Glaciers Melt Away." *Independent* (UK), November 18, 2006.

Connor, Steve. 2003. "Global Warming May Wipe Out a Fifth of Wild Flower Species, Study Warns." *Independent UK,* June 17.

Coral Reefs and Global Climate Change: Potential Contributions of Climate Change to Stresses on Coral Reef Ecosystems. Arlington, VA.: Pew Center on Global Climate Change.

Costanza, Robert, Ralph d'Arge, Rudolf de Groot, Stephen Farber, Monica Grasso, Bruce Hannon, Karin Limburg, Shahid Naeem, R. V. O'Neill, J. Paruelo, R. G. Rakin, P. Sutton, and M. van den Belt. 1997. "The Value of the World's Ecosystem Services and Natural Capital." *Nature,* 387–6230, 253–261.

Daily, Gretchen. 1997. *Nature's Services: Societal Dependence on Natural Ecosystems.* Washington, D.C.: Island Press.

Daily, Gretchen, and Katherine Ellison. 2002. *The New Economy of Nature.* Washington, D.C.: Island Press.

Dalton, Kathleen. 2002. *Theodore Roosevelt: A Strenuous Life.* New York: Vintage.

Daly, Herman, and John Cobb, Jr. 1989. *For the Common Good.* Boston: Beacon Press.

Daly, Herman, and Joshua Farley. 2004. *Ecological Economics.* Washington, D.C.: Island Press.

Deneen, Sally, and Tracey C. Rembert. 1999. "Uprooted: The Worldwide Plant Crisis is Accelerating." *E Magazine* (July–August) X-4.

REFERENCES

Dionne, E. J. 2004. *Stand Up Fight Back: Republican Toughs, Democratic Wimps, and the Politics of Revenge.* New York: Simon and Schuster.

Environmental Defense. 2004. Scorecard web site, Animal Waste, http://www.scorecard.org/env-releases/def/aw_gen.html, accessed January 5, 2005.

Erwin, J. M. et al. 2001. "The Great Ape Aging Project: A Resource for Comparative Study of Behavior, Cognition, Health and Neurobiology." In *All Apes Great and Small,* volume 1. *African Apes,* ed. Birute M. F. Galdikas et al. New York: Kluwer.

Frank, Thomas. 2004. *What's the Matter with Kansas? How Conservatives Won the Heart of America.* New York: Henry Holt.

Galdikas, Birute M. F., et al., eds. *All Apes Great and Small,* volume 1. *African Apes.* New York: Kluwer.

Gelbspan, Ross. 2004. *Boiling Point.* New York: Basic Books.

Globe and Mail. 2004. "A Warm Welcome to the Attitude Change on Global Climate Change." Toronto *Globe and Mail,* July 8, 2004.

Goodstein, Eban. 2004. *Economics and the Environment.* New York: Wiley.

Goodstein, Eban, and Justin Phillips. 2000. "Growth Management and Housing Prices: The Case of Portland, OR." Contemporary Economic Policy, 18–3, 334–44.

Goodstein, Laurie. 2005. "Evangelical Leaders Swing Influence Behind Effort to Combat Global Warming." *New York Times,* March 10, 2005, A1.

Gorman, James. 2004. "Side Effects; No Skeeters, No Problem? Not So Fast." *New York Times,* June 22, 2004.

Hall, Myrna H. P., and Daniel B. Fagre. 2003. "Modeled Climate-Induced Glacier Change in Glacier National Park, 1850–2100." *BioScience* 53 No. 2 (February).

Hakim, Danny. 2004. "California Backs Plan for Big Cut in Car Emissions." *New York Times,* September 25, 2005.

Hamer, Dean. 2004. *The God Gene: How Faith is Hardwired into Our Genes.* New York: Doubleday.

Hansen, James. 2006. "Climate Change: On the Edge." *Independent* (UK), Friday, February 17, 2006.

Hansen, James. 2005. "A Slippery Slope: How Much Global Warming Constitutes 'dangerous anthropogenic interference?'" *Climatic Change.* 68, 269–279.

Heal, Geoffrey. *Nature and the Marketplace: Capturing the Value of the Ecosystem.* Washington, D.C.: Island Press, 2000.

Hawken, Paul. 1994. *The Ecology of Commerce.* San Francisco: Harper Collins.

Hill, Julia Butterfly, and Jessica Hurley. 2002. *One Makes the Difference.* San Francisco: Harper Collins.

Hirsch, Tim. 2003. "Grim Future for Gorillas and Chimps." BBC News Service, April 6, http://news.bbc.co.uk/2/hi/science/nature/2921669.stm, accessed January 2, 2005.

Hume, David. 1966. *An Enquiry Concerning the Principles of Morals.* Lasalle, Ill.: Open Court Press.

Imperial College. 2003. *Crop Diversity at Risk: The Case for Sustaining Crop Collections.* Wye, UK: Department of Agriculture, Imperial College.

IPCC. 2001. *Summary for Policy Makers.* New York: UN Intergovernmental Panel on Climate Change.

Judson, Olivia. 2002. *Dr. Tatiana's Sex Advice to All Creation.* New York: Owl Books.

Kahn, Peter. 1999. *The Human Relationship with Nature: Development and Culture.* Cambridge, Mass.: MIT Press.

Keynes, John Maynard. 1972. "Economic Possibilities for Our Grandchildren." In *The Collected Writings of John Maynard Keynes,* vol. 9. London: Macmillan.

Kareiva, Peter, and Simon Levin. 2003. *The Importance of Species: Perspectives on Expendibility and Triage.* Princeton: Princeton University Press.

Kirby, Alex. 2003. "Alert Sounds for Turtles and Tortoises." BBC News Service, May 14, http://news.bbc.co.uk/2/hi/science/nature/3024643.stm, accessed January 2, 2005.

Kirkpatrick, L. A. 1999. "Toward an Evolutionary Psychology of Religion and Personality." *Journal of Personality,* 67, 921–952.

Klein, Rick. 2006. "Kennedy Faces Fight on Cape Wind [*Periodical Title*] April 27, 2006, A1.

Koomey, John, R. Cooper Richey, Skip Laitner, Alan Sanstad, Robert Markel, and Chris Murray. 1998. "Technology and Greenhouse Gas Emissions: An Integrated Scenario Analysis Using the LBNL-NEMS Model." Berkeley: Lawrence Berkeley National Laboratory.

Lichatowich, Jim. 2001. *Salmon Without Rivers.* Washington, D.C.: Island Press.

Lochner, Janis, Mary Kingma, Samuel Kuhn, C. Daniel Meliza, Bryan Cutler, and Bethe A. Scalettar. 1998. "Real Time Imaging of the Axonal Transport of Granules Containing a Tissue Plasminogen Activator/ Green Fluorescent Protein Hybrid." *Molecular Biology of the Cell,* no. 9: 2463–76.

Locke, John. 1980. *Second Treatise of Government.* Indianapolis: Hackett.

Loomis, John B. 1996. "Measuring the Economic Benefits of Removing Dams and Restoring the Elwha River: Results of a Contingent Valuation Survey." *Water Resources Research* 32, no. 2: 441–47.

Lovelock, James. "The Revenge of Gaia: Why the Earth Is Fighting Back— and How We Can Still Save Humanity." *Perseus,* 2006.

Maddison, Angus. 1995. *Monitoring the World Economy, 1820–1992.* Paris: OECD.

McCarthy, Michael. 2005. "Greenhouse Gas Threatens Marine Life." *The Independent,* February 4.

———. 2004. "Disaster at Sea: Global Warming Hits UK. *Independent, UK,* July 30.

McKibben, Bill. 1999. "Climate Change and the Unraveling of Creation." *Christian Century* (December 8): 116–34, 1196–99.

MacPhee, Ross, and Hans-Dieter Sues, ed. 1999. *Extinctions in Near Time: Causes, Contexts and Consequences.* New York: Kluwer.

Monroe, Bill. 2004. "Global Warming Giving Pacific Northwest's Weather that California Feel." *The Portland Oregonian,* May 16, 2004.

Morrison, Jim. 2004. "The Incredible Shrinking Polar Bears." National Wildlife, Feb/Mar. 42-2.

National Academy of Sciences. 2002. Effectiveness and Impact of Corporate Average Fuel Economy (CAFE) Standards. Washington, D.C.: National Academy Press.

Nelson, Richard K. 1993. "Searching for the Lost Arrow: Physical and Spritual Ecology in the Hunter's World." In *The Biophilia Hypothesis,* ed. S. R. Kellert and E. O. Wilson. Washington, D.C.: Island Press.

———. 1986. *Make Prayers to the Raven: A Koyukon View of the Northern Forest.* Chicago: University of Chicago Press.

Nickerson, Colin. 2001. "An Early Melting Hurts Seals, Hunters in Canada." *Boston Globe,* April 1, 2001.

Norton, Bryan. 1987. *Why Preserve Natural Variety?* (Princeton: Princeton University Press.

O'Neill, B. C., and M. Oppenheimer. 2002. "Dangerous Climate Impacts and the Kyoto Protocol." *Science,* 296 (June 14) 1971–72.

Orr, James C., Victoria J. Fabry, Olivier Aumont, Laurent Bopp, Scott C. Doney, Richard A. Feely, Anand Gnanadesikan, et al. 2005. "Anthropogenic Ocean Acidification over the Twenty-first Century and Its Impact on Calcifying Organisms." *Nature* 437 (29 September).

Patterson, F. G. P., and M. L. Matevia. 2001. "The Status of Gorillas Worldwide." In *All Apes Great and Small,* volume 1. *African Apes,* ed. Birute M. F. Galdikas et al. New York: Kluwer.

Patterson, F. G. P., and W. Gordon. 2001. "Twenty-seven Years of Project Koko and Michael." In *All Apes Great and Small,* volume 1. *African Apes,* ed. Birute M. F. Galdikas et al. New York: Kluwer.

Paulson, Tom. 2004. "Dogs May Hold Key to Unlock Human Diseases." *Seattle Post-Intelligencer,* May 21, 2004.

Podger, Corrine. 2002. "Quarter of Mammals 'Face Extinction'" BBC News

Service, May 21, 2002, http://news.bbc.co.uk/2/hi/science/nature/ 2000325.stm, accessed January 3, 2005.

Polasky, Steven, Charles Costello, and Andrew Solow. 2005a. "The Economics of Biodiversity." In *The Handbook of Environmental Economics*, ed. J. Vincent and K. G. Maler. Amsterdam: Elsevier.

Polasky, Stephen, Erik Nelson, Eric Lonsdorf, Paul Fackler, Anthony Starfield. 2005. "Conserving Species in a Working Landscape: Land Use with Biological and Economic Objectives." *Ecological Applications*, 15, no.4 (2005) 1387–1401.

Praded, Joni. 2000. "The Heat Is On: Global-Warming and Polar Bears." Animals (July 1): [pp].

Ralston, Holmes, III. 1994. *Conserving Natural Value*. New York: Columbia University Press.

Rockwell, David. 1991. *Giving Voice to Bear: North American Indian Myths, Rituals, and Images of the Bear*. Lanham, Md.: Roberts Rinehart.

Sagoff, Mark. 2005. "Carrying Capacity and Ecological Economics." *Bioscience* 45, no. 9: 610–19.

Scherer, Glenn. 2004. "The Godly Must be Crazy." *Grist Magazine*, November 4, 2004.

Scully, Malcolm. 2005. "An Unsettled Forecast for Global Warming." *Chronicle of Higher Education*, January 7, 2005, B12.

Slivinski, Stephen. 2004. "The Corporate Welfare Budget: Bigger Than Ever." Cato Policy Analysis no. 415. Washington, D.C.: Cato Institute.

Spotts, Peter. 2004. "An Arctic Alert on Global Warming." *Christian Science Monitor*, November 9, 2004. www.csmonitor.com/2004/1109/p01s03-sten.html.

Stern, Sir Nicholas. 2006. Stern Review on the Economics of Climate Change. London: UK Government, Department of the Treasury.

Sunstein, Cass. 2006. "It's Only $300 Billion" *Washington Post* Wednesday, May 10, 2006, A25.

Suzuki Foundation. 2005. "Who's Meeting their Kyoto Targets?" David Suzuki Foundation: http://www.davidsuzuki.org/files/climate/cop/ Meeting_Kyoto_Targets.pdf.

Swope, Christopher. 2006. "Coal Converts." *Governing*, 19, no. 7 (April), 44–48. Langer, Gary, "Public Concern on Warming Gains Intensity" March 26, 2006. ABC News: http://abcnews.go.com/Technology/ GlobalWarming/story?id=1750492&page=1.

Teriault, Carmen. 2002. "Plants That Heal the Earth." *American Gardener* (November–December).

Thins, J. K. H., and J. P. Evans. 1997. "Effects of Anthracnose on Dogwood Mortality and Forest Composition of the Cumberland

Plateau (U.S.A.)." *Conservation Biology* 11, no. 6 (December): 1430–35.

Thomas, Chris D., Alison Cameron, Rhys E. Green, Michel Bakkenes, Linda J. Beaumont, Yvonne C. Collingham, Barend F. N. Erasmus, Marinez Ferreira de Siqueira, et al. 2004. "Extinction Risk from Climate Change." *Nature* 427 (January 8), 145–148.

Tooby, John, and Liba Cosmides. 1992. "The Psychological Foundation of Culture." In *The Adapted Mind: Evolutionary Psychology and the Generation of Culture*, ed. Jeroem Barkos, John Tooby, and Liba Cosmides. New York: Oxford University Press.

Toronto Star. 2004. "Farmers Grab Gorilla Parkland." *Toronto Star*, July 16, 2004.

Union of Concerned Scientists. 2006. "Bush Administration FY07 Budget—Highlights and Lowlights." UCS web site, http://www.ucsusa.org/news/positions/president-bushs-fy-2007-budget.html, accessed August 13, 2006.

U.S. Energy Information Administration. 2006. "U.S. Carbon Dioxide Emissions from Energy Sources 2005 Flash Estimate." U.S. EIA web site http://www.eia.doe.gov/oiaf/1605/flash/flash.html, accessed September 2, 2006.

U.S. EPA. 2002. "U.S. EPA Wastewater Technology Fact Sheet: The Living Machine." http://www.epa.gov/owm/mtb/living_machine.pdf, accessed January 5, 2005.

USGS. 2004. "Where Have All the Frogs Gone?" U.S. Geological Survey web site, http://www.usgs.gov/amphibian_faq.html, accessed January 2, 2005.

WHO. 2004. "Environmental Risks." In *The World Health Report*, chapter 4. World Health Organization web site, http://www.who.int/whr/2002/chapter4/en/index7.html, accessed January 5, 2005.

Williams, Dennis. 2002. *God's Wilds*. College Station: Texas A&M.

Williams, Terry Tempest. 2004. *The Open Spaces of Democracy*. Great Barrington, Mass.: Orion Society.

Wilson, Edward O. 2002. *The Future of Life*. New York: Knopf.

———. 1992. *The Diversity of Life*. Cambridge, Mass.: Harvard University Press.

———. 1984. *Biophilia*. Cambridge, Mass.: Harvard University Press.

Wolf, Edward, ed. 2005. *The Economic Impacts of Climate Change in Oregon: A Preliminary Analysis by Economists*. Eugene: Resource Innovations, University of Oregon.

World Commission on Environment and Development. 1987. *Our Common Future*. New York: Oxford University Press.

Wrangham, Richard, and D. Pilbeam. 2001. "African Apes as Time Machines." in *All Apes Great and Small,* volume 1. *African Apes,* ed. Birute M. F. Galdikas. New York: Kluwer.

Young, Emma. 2003. "Great Barrier Reef to Be Decimated by 2050." newSceintist.com News Service, February 23, 2003, http://www .newscientist.com/article.ns?id=dn4707, accessed January 2, 2005.

INDEX

161